RETAILING FARM AND LIGHT INDUSTRIAL EQUIPMENT

RETAILING FARM AND LIGHT INDUSTRIAL EQUIPMENT

MANSEL M. MAYEUX

Professor
Agricultural Engineering Department
Louisiana State University
Baton Rouge, Louisiana

AVI Publishing Company, Inc.
WESTPORT, CONNECTICUT

Library of Congress Cataloging in Publication Data

Mayeux, Mansel M.
  Retailing farm and light industrial equipment.

  Bibliography: p.
  Includes index.
  1. Selling—Farm equipment. 2. Selling—
Industrial equipment. I. Title.
HF5439.F37M39 1983   681′.763′0688   83-2644
ISBN 0-87055-414-X

Cover: top right photo courtesy Grant Heilman Photography

Contents

Acknowledgments

The writer is indebted to the National Farm and Power Equipment Dealers Association for the use of its cost and management studies; to International Harvester Company, The John Deere Company, the Ford Motor Company, the J.I. Case Company, and White Motor Corporation for various trade information furnished; to Jimmy Dillon and Russel Mayeux, Deep South Farm and Implement Company; Jim Breeden, Breeden Tractor and Company, and Norman Lovitt, Lovitt Equipment Company, for their suggestions and review of the manuscript; and finally to Gayle Gautreau for her patience and care in typing the manuscript.

Related AVI Books

Modern Farm Power and Its Distribution

Agriculture is a vital cog in the American way of life. Its mechanization is mandatory if we are to maintain the standard of living we have become accustomed to. It is in the sales and service of the machinery that this industry uses that the farm equipment retailer plays the leading role.

Recent data released by the U.S. Department of Agriculture emphasize the importance of agriculture to our total economy and in particular to rural communities. Even with the industrialization of many rural communities, agriculture continues to be the backbone of most of these communities.

Farm operator families spend about $188.6 billion dollars a year for goods and services to produce crops and livestock. In 1979, they had available an additional $31.0 billion of net farm income and $34.0 billion from off-farm sources to spend for taxes, investments, and a few of the good things of life. In reality, the goods and services bought by farmers keep our national economy rolling. The USDA release gave the following data for 1977:

Farmers' annual purchases include:

—$10.9 billion worth of new farm tractors and other motor vehicles, machinery, and equipment. It takes 140,000 employees to produce this farm equipment.

—$11.7 billion for fuel, lubricants and maintenance of machinery and motor vehicles. Farming uses more petroleum than any other single industry.

—$17.4 billion for feed and seed.

—$6.9 billion for fertilizer and lime.

—360 million pounds of rubber in various products. This is about five percent of the total rubber used in the United States, and would be enough to put tires on seven million automobiles.

—33 billion kilowatt hours of electricity. This is equal to two percent of the nation's total, and is more than the annual residential use in all of New England plus Maryland, Kentucky, and Washington, D.C.

—6½ million tons of steel in the form of farm machinery, trucks, cars, fencing and building materials. Farm use of steel supports 40,000 workers in the steel industry.

In fact, the things farmers buy, just for production purposes alone, create jobs for two million people. Another eight to 10 million people have jobs storing, transporting, processing and merchandising agricultural products. These include over 308,000 employees and a payroll of $3.2 billion to workers engaged in meat-packing and poultry processing; 170,000 workers and a $1.8 billion payroll in the dairy industry; 237,000 workers and a payroll of $2.2 billion in the baking industry; 223,000 workers and a $1.7 billion payroll in canned goods and frozen foods industry; and 152,000 workers and a billion dollar payroll in the textile industry.

Together, this adds up to one out of every five jobs in private industry.

The average investment in farmland, livestock, machinery, and other assets was $400,000 per farm in 1980. This is practically double that of 1975 and triple that of 1970. Farm indebtedness grew even faster, averaging $68,200 per farm in 1980, up from $32,900 in 1975 and $18,000 in 1970. The rise in value of farm real estate has kept pace making it an important factor in the total wealth of the farming sector.

The size of farms is a major factor in equipment sales. In 1979, 56.6% of farms sold less than $20,000 worth of farm products, 14.0% sold between $20,000 and $40,000, 19.2% between $40,000 and $100,000, and 6.4% between $100,000 and $200,000. Only 3.3% of our farms sell over $200,000 of products annually.

It is abundantly clear that when agriculture prospers the entire economy prospers.

THE PROSPECT

Changes on the farm front have been so rapid that they really can be termed a revolution. Efficiency has been and will continue to be the key word. Equipment retailers must be prepared to contribute to this efficiency.

In spite of this revolution, farm machinery has remained relatively stable compared to the other inputs to agricultural production.

The year 1974 will probably go down in history as the year of the "Energy Crunch." It is real and a vital force in the decisions that farmers and equipment retailers make. More than ever, retailers must become "power specialists." They must be prepared to advise farmers on their power needs. To the extent that this advice is sound, dealers will prosper.

Mechanized farming is expensive in terms of energy consumption. Using 1967 as a base, diesel fuel has increased 410%; L. P. gas, 305%; gasoline, 290%; labor, 250%; and electricity, 195%. Table 1.1 shows

TABLE 1.1. AVERAGE ENERGY INPUTS PER HECTARE IN CORN PRODUCTION DURING DIFFERENT YEARS

Inputs	1945	1950	1954	1959	1964	1970
Labor (hours)[1]	57	44	42	35	27	22
Machinery (kcal $\times 10^3$)[2]	445	518	741	865	1038	1038
Gasoline (liters)	141	163	178	188	195	205
Nitrogen (kg)	7.9	16.8	30.1	45.9	65.0	150
Phosphorus (kg)	7.9	11	13	18	20	35
Potassium (kg)	5.7	11	20	34	33	67
Seeds for planting (bushels)	0.42	0.49	0.61	0.74	0.82	0.82
Irrigation (kcal $\times 10^3$)	47	57	67	77	84	84
Insecticides (kg)	0	0.1	0.3	0.8	1.1	1.1
Herbicides (kg)	0	0.05	0.1	0.3	0.4	1.1
Drying (kcal $\times 10^3$)	25	74	148	247	296	296
Electricity (kcal $\times 10^3$)	79	133	247	345	501	766
Transportation (kcal $\times 10^3$)	49	74	111	148	173	173
Corn yields (bushels)	84	94	101	133	168	200

[1] Mean hours of labor per crop ha in United States. Reported by David Pimentel, New York State College of Agriculture and Life Science, Cornell University, Ithaca, New York.

[2] An estimate of the energy inputs for the construction and repair of tractors, trucks, and other farm machinery was obtained from the data of Berry (R. S. Berry, "The Energy Cost of Automobiles," Science and Public Affairs, December 1973, pp. 58–60), who calculated that about 31,968,000 kcal of energy was necessary to construct an average automobile weighing about 1542 kg. In our calculations, we assumed that 244,555,000 kcal (an equivalent of 13 tons of machinery) were used for the production of all machinery (tractors, trucks, and miscellaneous) to farm 25 ha of corn. This machinery was assumed to function for 10 years. Repairs 15,000,000 kcal. Hence, a conservative estimate for the production and repair of farm machinery per corn ha per year for 1970 was 1,037,839 kcal.

the average energy inputs in corn production as computed by Dr. David Pimentel and others of New York State College of Agriculture and Life Sciences.

It is interesting to change these inputs and express them in terms of kilocalories. Over the years, corn yields have more than doubled. As we mechanized our production, the amount of fossil energy required to produce an acre increased tremendously. The amount of solar energy used by the plants remained relatively constant. The end result is that we are using more energy now in comparison to the results we are getting than we obtained prior to full scale mechanization (see Table 1.2). Notice the decline in labor inputs.

TABLE 1.2.   ENERGY INPUTS (KILOCALORIES PER ACRE) IN CORN PRODUCTION

Input	1945	1950	1954	1959	1964	1970
Labor[1]	12,500	9,800	9,300	7,600	6,000	4,900
Machinery[2]	180,000	250,000	300,000	350,000	420,000	420,000
Gasoline[3]	543,400	615,800	688,300	724,500	760,700	797,000
Nitrogen[4]	58,800	126,000	226,800	344,400	487,200	940,800
Phosphorus[5]	10,600	15,200	18,200	24,300	27,400	47,100
Potassium[6]	5,200	10,500	50,400	60,400	68,000	58,000
Seeds for planting[7]	34,000	40,400	18,900	36,500	30,400	63,000
Irrigation[2]	19,000	23,000	27,000	31,000	34,000	34,000
Insecticides[8]	0	1,100	3,300	7,700	11,000	11,000
Herbicides[9]	0	600	1,100	2,800	4,200	11,000
Drying[2]	10,000	30,000	60,000	100,000	120,000	120,000
Electricity[2]	32,000	54,000	100,000	140,000	203,000	310,000
Transportation[2]	20,000	30,000	45,000	60,000	70,000	70,000
TOTAL INPUTS	925,500	1,206,400	1,548,300	1,889,200	2,241,900	2,896,800
Corn yield (output)[10]	3,427,200	3,830,400	4,132,800	5,443,200	6,854,400	8,165,800
kcal return/input kcal	3.70	3.18	2.67	2.88	3.06	2.82

From the literature by David Pimentel, New York State College of Agriculture and Life Science, Cornell University, Ithaca, New York.
[1] It is assumed that a farm laborer consumes 21,770 kcal per week and works a 40-hour week. For 1970: (9 hours/40 hours) × 21,770 kcal = 4900 kcal.
[2] See Table 1.1.
[3] Gasoline, 1 gallon = 36,225 kcal.
[4] Nitrogen, 1 pound = 8400 kcal, including production and processing.
[5] Phosphorus, 1 pound = 1520 kcal, including mining and processing.
[6] Potassium, 1 pound = 1050 kcal, including mining and processing.
[7] Corn seed, 1 pound = 1800 kcal. This energy input was doubled because of the effort employed in producing hybrid seed corn.
[8] Insecticides, 1 pound = 11,000 kcal including production and processing (similar to herbicide; see[9]).
[9] Herbicides, 1 pound = 11,000 kcal including production and processing.
[10] Each pound of corn was assumed to contain 1800 kcal and a bushel of corn was considered to be 56 pounds.

Predicting the future is always dangerous. A certain amount of this is necessary. What can we anticipate in the next 10 years?

1. The rate of labor decrease will slow down and labor will become more skilled.
2. The size of farms will continue to creep upwards.
3. Farm managers will be better informed.
4. The rapidly spiraling power trend will be arrested.
5. Further urbanization of the farm population.

Some things will change very little in the future. To be successful as a retailer you will have to:

1. Have faith in your product.
2. Know its value to the buyer.
3. Know how it serves his needs.
4. Know how it can increase his production.
5. Know how it increases his leisure time.
6. Know how it can increase his standard of living.

7. Be able to communicate.
8. Discipline yourself to work long hours.
9. Discipline yourself to make the management decisions that your bookkeeping indicates.
10. Keep up to date in business, agricultural, and mechanization trends.

METHODS OF DISTRIBUTION

Over the years, there has evolved a time-tested method of distribution from factory to wholesaler or branch house to retailer. Each of these steps is vital to the industry. Though the ownership of these elements of the distribution chain may change, the functions they perform are not likely to change.

Functions of Manufacturer

The manufacturer has the primary responsibility for (1) research and development of equipment. This requires a great deal of coordination with many agencies. He must keep current with farming trends and practices. He must maintain liaison with universities and other public agencies doing research in agriculture and agricultural machinery. He must keep current in manufacturing techniques and materials. In short, his staff must constantly have the know-how to produce today the machines farmers will want tomorrow.

The manufacturer must (2) build functional equipment. Manufacturing facilities and techniques must be kept modern. Quality control must be ever-active to see that equipment that leaves the retailers' shelves can perform its intended function. Though the retailer has the responsibility of maintaining and servicing the equipment the manufacturer must (3) conduct training schools for mechanics to introduce them to the new machinery being produced. He must further (4) produce the parts necessary to service this equipment over its lifetime.

This machinery and parts must be (5) distributed to branch houses and wholesalers around the trade area. This entails a transportation function. The manufacturer must also (6) select and maintain effective branch houses over the trade area. The functions of the manufacturer are many and could be detailed to a greater degree but for the purpose of this text the six functions listed should suffice.

1. Research and development
2. Production of functional equipment
3. Conduct of service schools
4. Production of repair parts
5. Selection and maintenance of effective branch houses
6. Distribution of equipment to branch houses and wholesalers

Functions of Wholesaler and Branch Houses

The branch house is an arm of the manufacturer. It differs from the wholesaler not in function, but in the fact that the wholesaler may and usually does represent several manufacturers. The branch house represents only one.

"How wholesalers assist manufacturers" has been aptly described by John R. Bromell, of the Marketing Division of the U.S. Department of Commerce. He outlines the following ways in which the wholesaler helps the manufacturer of the finished product, and practically all of the functions apply equally well to distribution through branch houses.

1. *Storage*—Many wholesalers provide storage space for the manufacturer. They buy and store goods at the time they are produced and hold them until they are needed by local retailers. This greatly reduces the amount of warehouse space needed by manufacturers and eliminates the need for storage space by many small producers.

This system is sound economically, for the wholesaler can make better utilization of storage space than can most manufacturers. The production of many manufacturers is seasonal, and were it not for the wholesaler they would have to provide storage space for their maximum seasonal output. Therefore, as the goods are moved out, space would be vacated which would remain idle until opening of the next season.

The wholesaler, on the other hand, carries goods for every season, and as goods for one season are moved out, those for the next season are moved in. Consequently, space is more fully utilized.

2. *Finances*—The wholesaler aids the manufacturer with his problems of finance. By buying and paying promptly for goods as they are produced, the wholesaler lessens substantially the amount of capital needed by the manufacturer. Capital requirements would be much larger if the manufacturer had to hold the goods until they are needed and paid for by retailers.

The buying and storing of goods as they are produced is of particular importance to small manufacturers of seasonal equipment who have limited resources. For example, there are makers of seasonal equipment who have no warehouses and whose capital is barely sufficient to procure the raw material and get the equipment ready for shipment. By distributing through wholesalers these manufacturers are able to ship their products in advance of the sales season. This gives them money with which to continue operations until the season is ended. It also eliminates the necessity for the small factory to have a storage warehouse, to maintain a year-round staff for shipping small orders as they come in from the various retailers, and to keep a selling staff of sufficient size to cover enough of the potential market to dispose of the entire production.

Though policy varies from one manufacturer to the other, there is a distinct trend indicating that equipment is no longer manufactured based on expected orders. The high cost of money, storage, and obsolescence dictates a no-stockpiling policy. Through a program of floor planning, the wholesaler assists the retailer in placing firm orders so that equipment will be manufactured and available.

3. *Sales and Distribution*—The wholesaler and branch house undertakes the task of sales and distribution for the manufacturer. Through him the factory can get national distribution more quickly and more thoroughly than by any other means. If the manufacturer attempts to sell direct to retailers he has to train and send into the field a large staff of salesmen who commonly enter upon their duties as strangers to their prospective retail customers.

4. *Protection of Market*—If there were no wholesalers, small and medium-sized independent retailers would constitute, at best, only a minor factor in our marketing scheme, and no manufacturer would enjoy sufficient coverage without selling through corporate chains, mail order houses, and cooperatives.

5. *Advice on Market Conditions*—As his distributive agent, the wholesaler or branch house should keep the manufacturer advised on market conditions, the nature of the goods which can most readily be sold in the market, the type of product which farmers seem to prefer, and the size of unit which makes the greatest appeal. Thus, where the manufacturer employs wholesalers or branches in many areas, he can receive a constant flow of marketing information that will guide him in making design and production decisions.

6. *Selection of Retail Agencies*—A manufacturer who desires to have the best retailer in an area to represent him must study the character and operation of all the retailers in an area. The wholesaler, through his salesmen, already has this information. If he is successful he knows intimately the operations, the character, and the financial resources of the retailers in his trade area.

7. *Lower Handling and Transportation Costs*—The wholesaler substantially lowers the handling cost of the manufacturers. When distributing through the wholesaler, the manufacturer often merely has to load full cars and ship them on. If he sold to retailers direct, it would be necessary for him to employ a much larger warehouse force to fill the multitude of small orders which, of necessity, would come in from retailers. Many of these small orders probably would not earn the lower carlot freight rates. This would necessitate special packing and weighing of many orders, with a consequent increase in expenses.

By ordering goods in carlots the wholesaler not only lowers the distribution costs of the manufacturer but he contributes to lower costs to retailers and consumers.

8. *Simplification of Credit*—By going through the wholesaler, the manufacturer greatly simplifies his credit problems. If he were selling direct to retailers, he would have to keep a check on hundreds or thousands of accounts in widely scattered localities. When the wholesaler checks the credit of a retailer he does so for all the lines he carries, which might run into the hundreds. Were it not for the wholesaler, there would have to be credit checks by several hundred manufacturers on each retailer. This would represent both actual cost to the manufacturers and a substantial economic waste to society.

Functions of the Retailer

It has been said that "Nothing happens until something is sold." Though this may not be literally true, sales on the local level govern manufacturing and distribution. The retailer is *the salesman*. His functions are certainly much broader than mere selling but all his functions support his selling effort.

1. *Stock suitable machinery.*   Of all the many pieces of equipment available, the local dealer shoulders the responsibility of what to stock. A basic function is to offer for sale in a community that equipment which will help producers to prosper. It must be on hand in the quantity and quality needed at a price farmers can afford to pay.

2. *Sell vigorously.*   Equipment and machinery on a lot is useless until it is sold. Retailers must fulfill the needs of their customers as

well as their wants. Needs are economic. Wants are desires—the desire to have the finest or newest or biggest. Both are acceptable reasons for selling.

3. *Service equipment.*   The finest equipment must be serviced periodically. Repairs must be made. Since the life of a machine is generally understood to be the length of time up to the point when the repairs equal the original cost, it can be seen that servicing equipment may be equal in magnitude to sales when parts are taken into consideration. It also accounts for the importance farmers place on dealers and dealers' service organization.

4. *Supply parts.*   The local dealer must stock parts for the equipment he sells. The wholesaler or parts depot is his backup but the retailer must answer directly to the customer.

5. *Arrange financing.*   Here the retailer has a twofold responsibility. He must finance or arrange for financing for his own inventory. Additionally, he must assist his customers in arranging for financing for their needs.

6. *Advise farmers.*   Retailers today must be specialists. They must be able to assess the needs of their customers and stock the equipment the customers should be buying. The proliferation of technology makes this task an important and complicated one.

7. *Feedback to distributor and manufacturers.*   Manufacturers conduct extensive field testing. This does not replace the "on the firing line" reaction customers have to new machinery. Retailers must assess these comments and relay them to the manufacturer. They must anticipate future trends and assist distributors in determining what will be sold. Some manufacturers have formalized a method of receiving this feedback. Additionally, it can be funneled in an organized way through the "trade association."

The manufacturer, the branch house or wholesaler, and the retailer perform vital functions. They mesh together as the cogs of a gear to keep the farmers' mechanized wheels rolling.

QUESTIONS

1. How does the "energy crunch" affect agricultural mechanization?
2. How will predictable trends affect the local retailer?
3. What is the manufacturer's most important function?
4. How does the "wholesaler" differ from the "branch house"?
5. What are the functions of the branch house or wholesaler?
6. What are the functions of the retailer?
7. What is the most important element in the distribution chain?

Starting a Farm Equipment Business

The average dealer in the United States in 1980, according to the "Cost of Doing Business Study" published by National Farm & Power Equipment Dealers Association, had a sales volume of $2,252,306, which netted an operating profit before income taxes of 2.87% of sales or $64,721. He had total assets of $1,238,677 of which $901,122 was inventories, $135,745 was in accounts receivable, and $117,148 was in fixed assets. His return on total assets was 5.22%. He employed approximately 17 employees, who were responsible for over $164,373 sales each. The low net profit and the high sales per employee makes management a key ingredient in the success equation.

Though these statistics represent the operation of the average dealer, as represented in the Cost of Doing Business Study, it must be understood that the dealers who participate in this study are probably above average. It must be further understood that in a country as big as the United States, with its very diverse agriculture, there are always depressed areas even in times of prosperity and conversely areas of affluence during generally depressed times.

THE DEALER

Success depends on the individual. Are you qualified, are you disciplined, are you willing to work, and do you have faith? To the experienced dealer many of the questions that puzzle the rookie may seem elementary. Several years ago the U.S. Department of Commerce published a booklet that points out many pertinent and searching questions most helpful to the beginner. They are not designed to deter anyone from starting a business, but rather to point out realistically some of the problems that might otherwise have to be learned wholly from experience.

Questions to be considered	Check to indicate that you have considered each point
1. Have you had previous experience in farm and power equipment or related lines?	( )
2. Have you ever bought merchandise for resale?	( )
3. Have you ever lived on a farm? Do you know the function of various kinds of equipment?	( )
4. Do you have technical skills needed in servicing tractors and other kinds of equipment?	( )
5. Have you ever supervised the work of others?	( )

6. Have you ever hired people and met payrolls? --------------------------------- (   )
7. Have you ever dealt with the public, and do you like to work with people? --- (   )
8. Have you ever sold either merchandise or service? --------------------------- (   )
9. Can you drive yourself to do what is necessary, be your own self-starter? -- (   )
10. Do you have imagination? Energy? Initiative? ------------------------------- (   )
11. Are you willing to work long hours? --------------------------------------- (   )
12. Can you overcome obstacles? Keep plugging? --------------------------------- (   )
13. Would working with someone else to gain experience be helpful? ------- (   )

FINANCIAL QUALIFICATIONS

14. How much cash have you saved to put into the business? ------------------ (   )
15. How much do you have in the form of other assets that could be sold and put into the business? --- (   )
16. Have you sources from which you could borrow money if needed? ------- (   )
17. Would it be advisable for you to save more money before starting? ----- (   )

ANTICIPATED INCOME

18. What are your present earnings? -- (   )
19. How much can you earn as the owner of the contemplated business? -- (   )
20. Is this more than you can make as an employee? --------------------------- (   )
21. Are you willing to be content with reduced earnings while getting started? --- (   )
22. Can you risk uncertain or irregular income for at least one year? ------ (   )
23. Are business conditions good or bad? --------------------------------------- (   )
24. How are conditions in the farm and power equipment field? -------------- (   )
25. Have any stores of this type closed recently? ------------------------------- (   )
26. Have you talked with a banker? What does he advise? --------------------- (   )
27. Would he be willing to loan money on your enterprise? -------------------- (   )
28. If you are buying a going business, why does the owner wish to sell? -- (   )
29. Have you checked his claims with men in nearby businesses of like kind? --- (   )
30. Has an attorney checked the purchase agreement for defects? ----------- (   )
31. How much of the inventory is current and salable? ------------------------- (   )
32. Are the office and store fixtures in good condition? ------------------------- (   )
33. Is your personal situation right for such a start? Your age? Your present job? Your future prospects? Your family responsibility? Your habits? --- (   )

Many other important phases are involved in a self-analysis of this kind, including business judgment. Most of these are predicated on the assumption that the main objective is to make a success of the business. A key question may well be, "What is your past record of handling financial matters?"

Attitude and the desire to achieve success are paramount.

THE TERRITORY

We hear stories about selling refrigerators to Eskimos but they are probably just that—stories. Determine the market potential of an area before starting your business. Base your decision on facts. First, determine the facts pertaining to the agriculture of the area. These include the crops and livestock grown, the farm sizes, the monetary returns to the farmer, and the trends in the farming community. The U.S. Census Bureau, personal surveys, the county agent, local USDA agencies, and Chambers of Commerce are all good sources. Determine the financial climate in the area. The condition of farm homes and farmsteads is an excellent indication of prosperity. Local financial institutions such as banks and production credit associations are excellent references. Friends in allied businesses and potential competitors can be of valuable assistance.

Determine the machinery needs of the proposed area. The U.S. Census is a good source to develop trends in machinery use. Personal surveys are mandatory in this area to determine up-to-the-minute existing conditions and to determine needs that are not being satisfied. Manufacturer's representatives who are experienced in territory analysis should be consulted. Your regional Farm and Power Equipment Dealers Association executive secretary can also contribute materially.

The U.S. Department of Commerce recommends that the following 14 questions be considered prior to making the final decision.

1. Are farmers in the locality prosperous? --- ( )
2. Are crops diversified, general farming, dairying, livestock? Or is it a "one crop" territory? --- ( )
3. Are the farm organizations, 4-H Clubs, Future Farmer Chapters, and Soil Conservation Districts, aggressive and adequate? ---------------------- ( )
4. How many stores in the territory are similar to the type you plan to operate? -- ( )
5. Are there good local outlets for farm products? Or are farmers compelled to market crops in other towns? -------------------------------------- ( )
6. Is the wealth and income of farmers fairly well distributed or must sales depend upon a few large operators of farms, ranches, or plantations? --- ( )
7. Are transportation facilities, professional services, banking facilities, schools, and the like good, adequate, or poor? --------------------------------- ( )
8. Do other farm equipment stores appear to be healthy and busy? -------- ( )
9. Are competitive stores alert and aggressive? Or is there complaint among farmers that they fail to render efficient service? ------------------- ( )
10. Are competitive stores old and well established? Modern and attractive? --- ( )
11. Have you talked with wholesalers and suppliers about prospects for an additional store in the community? --- ( )
12. Do other local businessmen in noncompeting lines give encouragement? --- ( )
13. If you are buying an established business, does it have the "good will" of the farming community? --- ( )
14. How many farms in your proposed territory? --------------------------------- ( )

TIME OF ENTRY

Entering the farm equipment business during a period of economic recovery and expansion is frequently the key to success. A knowledge and awareness of the four phases of a business cycle may serve as a guide in deciding upon the most favorable time to enter.

The four phases of a business cycle consist of expansion, peak, contraction and depression. It has been said that the length of a cycle, from peak to peak, is about 9 years.[1] Other authors recognize economic fluctuations and irregularities which pass through the phases mentioned above, but the sequence of events is not always so clearly defined.[2] There may be hesitations and reversals within each phase and at varying rates of change. There may be a long period

---

[1] Tarshis, L., The Elements of Economics, Houghton Mifflin Company, New York, 1947, p. 443.

[2] The irregularity and questionable nature of business cycles is aptly discussed by Kenneth Boulding in his revised edition of Economic Analysis, Harper and Brothers, New York, 1948, pp. 375–395.

of stability within any one phase and an almost imperceptible easing into another phase. Or there may be sudden adjustments plunging the nation from prosperity to depression within a very short time. Also, sudden periods of recovery as witnessed in times of national emergency are usually followed by prolonged periods of economic expansion.

The best time to start in business is difficult to predict, but one cannot refute the good economic judgment of avoiding heavy investment in a declining market. If business has been at its peak for a prolonged period of time, there is a strong probability that the contraction phase, characterized by declining prices, lower farm incomes, and reduced profits, will soon occur. On the other hand, a dealer who starts in business at the beginning of the expansion phase will find his chance for success greatly enhanced.

The period following World War II was characterized by an expansion of economic activity and proved to be an ideal time to enter the retail farm equipment business.

One out of every four of the firms doing business in 1962 had entered during the period 1945–1949. The net profit margin then was 7–10% of sales. Current profit margins are closer to 3% of sales. The year 1974 was another year of expanded sales. Profits rose to an average of 5.97% of sales, then declined steadily through 1977.

OBTAINING A FRANCHISE

Prospective dealers may obtain a franchise to sell farm equipment by negotiating with a representative of an equipment manufacturer for a Dealer's Sale Agreement or a Company Dealer Contract. The dealer applies to the farm equipment manufacturer of his choice. Conversely, it is not uncommon for the manufacturer to instigate a search for a competent individual to contract for a franchise. Regardless of who initiates the action, the prospective dealer must have the ability to buy, sell, and service equipment. The area in question must have potential for sales.

Some companies may require a guarantee bond to cover purchase of inventories, fixed assets, and miscellaneous start-up costs. Others may negotiate a title retention agreement, which stipulates the company's ownership and control of its goods.

Federal law prohibits the assignment of specific sales territories. The logistics of delivery and service, however, tend to restrict dealers to definite areas. A dealer is expected to develop the full sales potential of his surrounding area. This area must be large enough to provide a profitable business. Marginal trade territories profit no one. They lead to distressed merchandise, which tends to disturb the pricing structure of a territory.

Dealers may negotiate with more than one manufacturer for a franchise. In actual practice, manufacturers frown on having competitive major brands under one roof. The reason for this policy is quite obvious. Not the least of these reasons is the difficulty of properly financing and representing a multiple-brand operation in one community.

Capital requirements to start a business will depend on the size and type of market. The average North American dealer in 1980 had $285,648 of working capital. Working capital does not include fixed assets such as land, buildings, and equipment not held for resale.

This average dealer had sales of $2,252,306. Capital requirements were therefore $1.00 for each $7.88 of sales. Lower-volume dealers whose average sale was $720,778 had working capital of $142,887 or $1.00 of capital for each $8.65 of sales. Though many dealers have started on a "shoestring," most manufacturers are not anxious to negotiate a franchise with limited financial resources.

Farm equipment dealers may finance their inventories by various methods. The larger, more expensive items of equipment are usually obtained through a floor plan arrangement with the manufacturer. The floor plans vary in detail with regard to due dates, repayment schedules, and discounts. Most floor plans allow dealers from 6 to 14 months to pay for their inventory of machinery. The high cost of money is creating a trend towards shortening this floor plan period. In all cases, full payment is required as soon as the equipment is sold to a customer.

A sample of farm equipment dealers reports a range in manufacturer trade discounts from 17.5 to 23.0%. The trade discount represents the reduction in list price to arrive at the dealer's purchase price.

When a dealer pays for a machine before the designated due date, many manufacturers allow a prepayment discount. This discount varies from 2 to 4%, depending on the time remaining before the final due date.

A few farm equipment manufacturers offer volume discounts to dealers. These discounts are based on a dealer's total net purchases within a prescribed time period and range from 3 to 6%.

Discounts are also offered by some major farm equipment manufacturers when cash payments for parts are made before the due dates. In addition, most farm equipment manufacturers grant pre-season stock order discounts ranging from 2 to 4%. Some manufacturers have less favorable terms. They may require that parts be paid for by the 15th of the month or future deliveries will be shipped C.O.D. Supply and demand as well as interest rates on borrowed capital all affect these trade policies.

Sales agreements may be terminated by either party by mutual agreement. Termination can also occur if dealer or manufacturer fails to abide by the terms of the contract. Through the National Farm and Power Equipment Dealers Association, committees of dealers work with individual manufacturers on terms of equitable contracts. The importance of understanding the terms of the contract prior to signing the agreement cannot be over emphasized.

FINANCING THE FRANCHISE

Dealers may finance their inventories through commercial banks or other financing agencies by use of a trust receipt, a chattel mortgage, or a conditional sales contract. A trust receipt is a written agreement between the bank, the dealer, and the manufacturer. The bank buys and takes title to the equipment from the manufacturer. At the same time, the dealer is entrusted with the equipment and authorized to sell or otherwise use the equipment as specified in the agreement. The dealer agrees to account to the bank for all transactions and proceeds relating to the financed equipment. In addition, the dealer signs a promissory note for the full value of the equipment plus interest charges. The lending agency holds title to the equipment until full payment is received.

A floor plan chattel mortgage is a written agreement between the dealer and a lending agency, wherein the dealer who has title to inventory equipment pledges this equipment as security for borrowed capital. The required care, maintenance, use, and final disposition in the event of default are generally specified in the mortgage agreement.

A conditional sales contract differs from the chattel mortgage in that the title is retained by the dealer as security when the equipment is sold. The bank or other financing agency becomes involved only when it purchases the dealer's interest in a contract or takes assignment of the dealer's interest as collateral for a loan.

TYPE OF OWNERSHIP

Decisions as to type of ownership will hinge upon the economic characteristics found in a single proprietorship, partnership, or corporation. The larger firms are generally incorporated and the smaller ones are individual proprietorships.

Each type of ownership has its advantages and disadvantages. The single proprietor, of course, has full liability and control, limited capital strength, and limited life. The partnership permits two or more individuals to pool their skills, resources, and liabilities. The corporation, as the most widely accepted type of ownership, provides for greater capital strength, limited liability, and perpetuity. More detailed information on the advantages and disadvantages of incorporation may be obtained by consulting a competent attorney.

Businessmen frequently ask whether there is a particular income tax advantage associated with each of the three types of ownership. Apparently there is no essential difference between the tax rate paid by the individual entrepreneur and a partner of a firm. But there are some unique features about the corporation tax that may provide tax savings, particularly for high income firms. The individual proprietor pays 14% of the first increment of taxable income. This is graduated upward with each increment of added income until a maximum of 50% is reached. The corporate rate, on the other hand, is 16% for the first $25,000, 19% for the next $25,000, 30% for the next $25,000, 40% for the next $25,000 and 46% for all over $100,000. The concept of the surtax is not currently employed. The rates will decrease slightly in 1983.

Tax laws have been designed so that small businesses may incorporate but still be taxed as individuals. Since tax laws are constantly changing and the American tax structure is complex, the advice of competent tax attorneys is a necessity for most business operators. Peculiarities of incorporation laws and recent adjustments in tax rates suggest that decisions be based on advice of a competent authority.

ESTIMATING VOLUME

After all the facts and figures have been determined, the problem of estimating the volume of business still remains. The Cost of Doing Business Study for 1980, referred to previously, can be of much help (see Table 2.1). The operating averages available by regions show the volume of sales by category as a percentage of total sales. This can be used as a guide for your proposed operation. If you can estimate, for example, the expected sales of new equipment, you can now predict the total volume of sales as well as the sales of used equipment, repair parts, and service labor.

As an example, let's assume you determine you can sell 12 medium-sized tractors (70 hp) and 15 large tractors (120 hp) for a total new tractor sales of $488,000. Allied cultivators, planters, plows and miscellaneous new equipment should run 25% of the above or a total of $122,100 or a grand total of $610,100. Referring to page 6 of the Cost of Doing Business Study, we find that the North American average (choose the appropriate regional figures) new equipment sales represent 52.24% of total sales. Let x be the total sales.

$$0.5224\,x = \$610,100$$

$$x = \frac{610,100}{0.5224}$$

$$x = \$1,167,879.00 \text{ Total Sales}$$

Therefore, total sales are predicted to be $1,167,879. Used equipment sales should be 17.7% of total sales or 0.177 × $1,167,879 = $206,948.16. Repair parts should be 17.5% of sales and service, 6.7% of sales, or $204,830 and $77,780 respectively.

TABLE 2.1.   NORTH AMERICAN OPERATING AVERAGES FOR ALL DEALERS, 1980

Account Description	Average	
	Amount	% Sales
Sales:		
New equipment	1,176,693	52.24
Used equipment	399,201	17.72
Total new and used:	1,575,894	69.96
Repair parts	395,027	17.54
Service labor	149,875	6.66
All other lines	105.248	4.67
Total sales	2,252,306	100
Gross margins:		
New equipment	111,333	10.08
Used equipment	28,163	7.05
Total new and used:	146,799	9,32
Repair parts	121,658	30.8
Service labor	54,681	36.48
All other lines	15,737	14.95
Total operating margin	346,541	15.39
Total expenses	335,556	14,89
Net profit on sales	10,975	0.49
Other income	53,746	2.38
Net operating profit or loss	64,721	2.87

Taken from 1980 Cost of Doing Business Study (see Appendix A).

This system of predicting volume can be used effectively if and only if a good estimate can be arrived at for new equipment and if you are operating in an area where your brand of merchandise is well established. These percentages obviously would not apply with a new dealership introducing a brand not known to the area. Where statewide percentage figures are available, they should be used. Success in business goes to those who make their business succeed. There are four good sources of advice that anyone might be wise in consulting— your banker, your accountant, your friends in the industry, and your Farm and Power Equipment Dealers Association.

QUESTIONS

1. How do you obtain a franchise to sell farm equipment?
2. What do manufacturers want to know about a potential retailer?
3. When must "floor planned" equipment be paid for?
4. Discuss the advantages and disadvantages of the three types of ownership.
5. What are the dangers of trying to start a business on a "shoestring?"
6. How do you analyze the potential of a territory?
7. What are the pitfalls of planning a business based on averages?

Capitalization and Financing

Money talks, or should we begin by saying that nothing is quieter than a bankrupt business. Study after study shows that underfinancing is one of the leading causes of business failures. The farm equipment retailing business is certainly no exception. It cannot exist long without adequate capital.

BUILDING CAPITAL

There are many definitions of capital. A simple one will suffice. Capital is the concrete physical goods and funds employed in a business. It is usually expressed in terms of dollars. More specifically, it is the cash, the receivables, reserves, inventories, and fixed assets such as buildings, trucks, tools, and office machines used in the business.

Capital may be classified according to its source. We term the monies we borrow *borrowed capital* and the monies and goods we personally put into a business *equity* or *proprietary capital*. The latter is created only by consuming less than you produce. The former is acquired from lending agencies.

Capital may also be classified according to its use. Capital that is tied up in fixed assets is normally referred to as fixed capital. A fixed asset is inventory not held for resale. Fixed assets consist of buildings, vehicles, office machines, and other goods needed to operate a business but not held for sale.

Working capital, on the other hand, is unencumbered capital available to conduct business. More specifically it is the difference between your current assets and your current liabilities. Figure 3.1 may serve to illustrate.

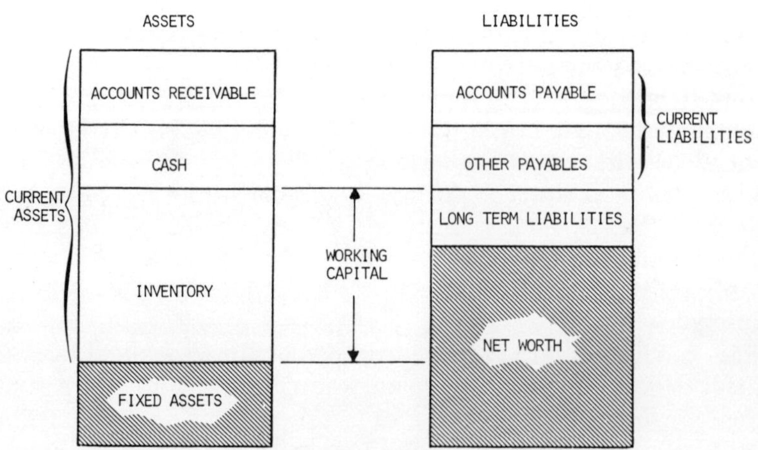

FIG. 3.1. Working capital

There are three ways in which you can increase your working capital:

1. Reduce your fixed assets
2. Increase your long-term liabilities
3. Increase your net worth

How can we reduce fixed assets? Lease or rent instead of buying. Sell your building and rent. Capital invested in real estate seldom returns more than 6–8% on the money invested. Lease trucks and cars used in the business. That expense then becomes a deductible cost of doing business. Rent or lease your major office machines. The effect of this on the working capital is readily understood by referring again to Fig. 3.1.

Increase your long-term liabilities. This may be easier said than done. To the extent possible, convert your accounts payable to long-term notes payable. Mortgage your property for cash. This would cause cash as well as long-term liabilities to go up.

The third way is to increase your net worth. Sell your inventory at a profit. This will cause your current assets to go up along with your net worth.

By definition, working capital is the difference between current assets and current liabilities. Short-term bank loans cannot increase working capital. They only increase payables and your cash, but have no effect on the working capital. Keep your working capital turning over. Establish a sound credit and collection policy. Refer customers who need long-term credit to suitable financial organizations.

Study the working capital cycle.

Remember that you have to have cash to get the wheel rolling. Buy suitable inventory and sell vigorously, then collect. A failure of any of the three steps will bring the wheel to a screeching halt.

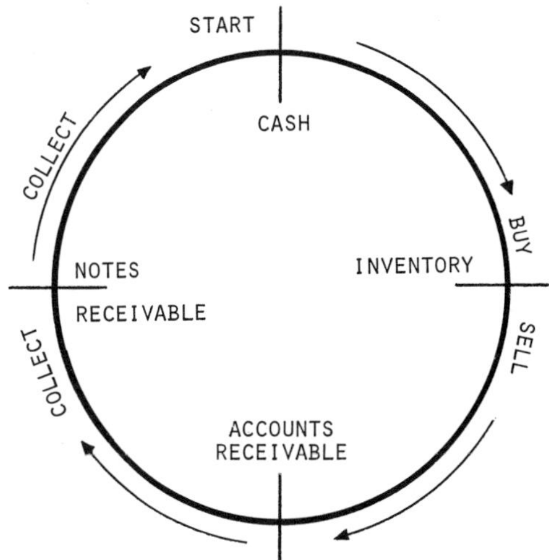

FIG. 3.2. Working capital cycle

MANAGING CAPITAL

Managing capital is seeing that cash, accounts and notes receivable, inventory, labor, and accounts payable are maintained in balance. It is difficult to say which is the most important. Each is necessary in its own way for the survival of the business. The important thing is balance.

Two financial ratios might be very helpful at this point. The first, known as the current ratio, is the ratio of current assets to current liabilities. It is computed by dividing your current assets by your current liabilities. It is an indication of your ability to pay out. You should strive for a 2 to 1 ratio. Ratios of 1.5 to 1 are generally acceptable, particularly with young dealers.

As important as the current ratio is the quick ratio. It is computed by dividing the cash and receivables by current liabilities. This shows you how liquid you are. If your creditors came calling could you pay off? A ratio of 0.5 to 1 is considered good. Many dealers operate with ratios as low as 0.2 to 1.

Since conditions vary so much from dealership to dealership, it is difficult to give guidance that would be applicable to all. The distribution of assets shown in Table 3.1 is taken from the 1980 Cost of Doing Business survey published by National Farm and Power Equipment Dealers Association for All North American Dealers.

It is evident from Table 3.1 that good managers balance their assets and keep a high percentage of their capital in cash. This enables them to take advantage of discounts and discourages distress sales.

TABLE 3.1. PERCENTAGE OF ASSETS BY CERTAIN CATEGORIES IN 1980: TYPICAL BALANCE SHEET FOR ALL NORTH AMERICAN DEALERS

Dealer	Cash and securities	Receivables	New equipment	Used	Parts	Other	Fixed assets
Average	3.76	10.96	44.39	11.95	11.99	3.30	9.48
High profit	5.61	13.39	42.02	10.01	11.29	2.81	9.38
Low profit	2.27	8.33	46.81	13.92	12.27	3.93	8.88

Remember that a business can go bankrupt while earning a profit just because of lack of cash. Bankruptcy is the inability to meet current obligations as they become due. The fact that you are making a profit is unimportant in this definition.

Notice how the high-profit dealer keeps his inventory down. Remember, it is not how much inventory you have on hand. The secret is having on hand what the customer wants. Some authorities make the blanket recommendation that used inventory should be kept below 15%. Notice that the high-profit dealer keeps his down below 11%.

OBTAINING CREDIT

A good money manager keeps good relations with his banker. He knows that the ability to obtain money when it is needed is of paramount importance. There are several sources of credit available to the dealer. Commercial banks, commercial credit companies, factoring companies, and certain governmental agencies should all be investigated.

The commercial bank is probably the leading source of credit. It is a service organization and is dependent on borrowers for its life blood. There is such a demand for its service that the banks can and indeed must be choosey. It is good for the dealer to understand how to do business with a bank.

The banker is obligated to keep certain firm obligations in mind. First, there are certain laws that govern the operation of the bank. These laws, along with regulations promulgated by banking commissions, spell out broad guidelines regarding collateral and reserves that banks must maintain. In addition, the obligation of the bank to its depositors and the demands that may be made on the bank by its depositors and creditors must be kept in mind at all times. For these reasons the banker must know all he can about the people to whom money is loaned. The dealer must make sure that he leaves a favorable impression on his banker.

How do you create this impression? By keeping the banker informed about your business and its condition not only when you need money but also when you are sitting in the driver's seat as well.

The Financial Advisory Committee of the Committee for Economic Development, an independent, private, nonprofit corporation, has distributed some helpful advice on banker-dealer relations.

Sit down and talk with your banker frequently; think out loud with him; give him your confidence; develop his confidence.

As his customer, you are entitled to open-minded energetic cooperation from him. There is no mystery about the loaning of money. It is his job to make loans—his opportunity to be of service. In serving you, he is rendering a constructive service to you, to his community and to his bank.

In working with your banker, bear in mind that his first obligation is to conduct a sound bank. He must finance business operations according to law and policies developed as a result of experience. The funds he lends you are largely the deposits in the bank, the composite funds of your community. Only a small part of a loan represents the bank's own funds. The obligation to the depositor must never be forgotten.

Plan Ahead

A good idea may remain in a desk drawer indefinitely for want of a way to finance it. And the soundest business plans may remain sterile unless sufficient capital and credit are provided—in advance of the time for action.

There is nothing mysterious about obtaining capital or credit, but it is essential to make plans ahead of time. Financial arrangements cannot be left to chance or afterthought; they require the same careful attention that is given to planning for production and sales.

Bear in mind that the techniques and procedures of finance change as does technology. New methods constantly are being developed to meet the changing needs of business.

No business program should be carried beyond its initial stage without being measured against the funds available in the business or against a reasonable estimate of the capital or credit obtainable elsewhere. In any case, estimates of required funds should not be based upon preconceived ideas or guesses, but on first-hand investigation.

On the other hand, intelligent foresight does not necessarily demand that every last detail be worked out in advance, or that one must have an assured source of capital or credit, before any other realistic planning is undertaken. Foresight simply means that these three financial considerations must be a continuing part of the business plans from the very beginning:

1) The financial structure of your business.
2) Financial controls in the daily operations of the business.
3) Availability of the funds that may be required.

Credit Based on Confidence

At the outset, your banker must have confidence in your integrity and your aims as a businessman—particularly your will to meet all financial obligations. He respects evidence of unrelenting effort to build up the financial strength of your business. He appreciates evidence of faith in the future of your company, as shown by the sacrifice of immediate income as an owner or major stockholder in order to strengthen your operating budget, thereby enlarging your capacity to earn in the future.

To assist him in appraising the efficiency of your organization, keep him fully informed. Furnish him with periodic balance sheets, income accounts, and surplus changes, and other specific information covering important developments in the company's financial or operating situation. If possible, supply him with an annual audit of your company's books by an outside firm of accountants. This audit gives both of you an unbiased report which measures operating efficiency. Discuss this report with your banker. Invite questions and suggestions.

When your management has demonstrated ability to make money under a variety of conditions, you have strong evidence of your capacity. However, a business that is going strong today may not need a long record of earnings or a strong financial statement in order to obtain bank credit. The most important single factor is conclusive evidence that your business is moving forward as a going concern. Your past record may be less important than present and future capacity, of which past performance is only partial evidence.

A capable banker never forgets that each industry has its own special characteristics. A financial statement and an income account which are good for a company in one line of business may be unsatisfactory in another field. A comparison of results for your business with those of other concerns in your own field is of equal importance to you as a businessman. Such analysis familiarizes you and your banker with financial conditions and practices in your trade which should be examined and discussed.

In this regard, the NFPEDA Annual Cost of Doing Business Study is very helpful.

Good Bookkeeping and Accounting

With adequate financial records and competent bookkeepers you can easily have the correct financial information which you and your banker require to keep abreast of affairs.

A good cost system with inventory control, always of great value, are of utmost importance in periods of rapid change or expansion. Knowing your cost—item by item, and department by department—you are much better prepared to reach decisions on expansion, and to give your banker a complete understanding of the soundness of such plans.

It is recognized that some organizations may not be large enough to support the specialized personnel required to keep such records. In such cases, outside accountants of proven ability can render this essential service. Periodic consultations with them are often productive of good results.

Keep in Touch

Continuing from the CED booklet,

Having established your banking connections, deal with your banker frankly on all business matters. Keep him fully informed. Discuss with him what you have done, what you are doing, what you propose to do, how and when.

It is to your advantage to meet with your banker from time to time, even though no immediate loan is in prospect. Talk with him informally about your plans. Invite his comments and suggestions. He may not have immediate answers to your problems, but he has many sources of information that can be of practical value to you and your business. When necessary, your banker can easily consult one or more of his correspondents to obtain specialized information of immediate value to you.

Do not feel that you are wasting his time or your own when you discuss your affairs with him. He is as much interested in studying your company's operations as you should be in keeping him informed on your progress and future plans.

Many bankers make a practice of visiting each customer's place of business from time to time. Encourage these visits and introduce your banker to your employees. Show him how your business is run. Make sure that he understands and appreciates the "know-how" of your management and your organization. Always present your situation in simple factual terms. Your banker respects you for recognizing problems that exist and developing sound ideas for solving them.

Managers Bulletin No. 11 of the National Retail Farm Equipment Association gives good advice to prospective borrowers and is quoted next.

Have the Answers Ready

Before a banker is prepared to make a loan he must feel satisfied with the answers to these five questions.

1. What sort of person are you, the prospective borrower?
2. What are you going to do with the money?
3. When and how do you plan to pay it back?
4. Is the "cushion" in the loan large enough? In other words, does the amount requested make suitable allowance for unexpected developments?
5. What is the outlook for you, the borrower, for your line of business in general?

What sort of person is the borrower?—The first question is by all odds the most important. The character of the borrower comes first. Next is his ability to run his business.

What about the use and repayment?—These points should be considered together. The banker will always want to know. What is the borrower going to do with the money and when is he going to pay it back?

What can you as a borrower do about preparing answers to the two questions? As a first step before you talk to your bankers about a loan, decide for yourself how much money you need, what you expect to do with it, and how you expect to pay it back. Then explain your plans to your banker, accurately, and in detail so that he may understand your program clearly.

Is there enough cushion in the loan?—is often the cause of honest differences of opinion. The borrower for his part may believe that there is enough cushion when, as the banker sees it, there is not.

PRESENTING FINANCIAL DATA

If the banker is satisfied with the answer to the first four questions, he can make a loan on the basis of:

1. Financial statement showing the condition of the borrower, or
2. Collateral pledged

If one of these two conditions is not met, the banker may run into well-merited criticism from the bank examiners. These examiners have a duty to perform in the interest of the bank's depositors and the public welfare.

Adequate figures a "must." When you plan to borrow money, you can do your part toward building an effective banking relationship if you will make available willingly to the banker:

1. Your balance sheet and profit-and-loss statements.
2. Other financial data where needed in sufficient detail. To illustrate: A financial report which shows the "aging" of accounts receivable is more informative than one which shows only the total amount.
3. Financial reports at sufficiently frequent intervals, so that the banker does not have to guess at what is going on in your business.
4. Balance sheets and profit-and-loss statements by acceptable certified public accountants.

Using collateral. If the loan required cannot be justified by the borrower's financial statements alone, a pledge of collateral may bridge the gap. If the collateral consists of readily marketable stocks and bonds, or the cash surrender value of a life insurance policy, the road is usually smooth. Values are easily established.

Other types of collateral, however, must be considered in a different light. Values are not always easily agreed upon. The law with respect to a valid pledge differs in different states. Moreover, what constitutes a safe margin can well be a matter of opinion, rather than a generally accepted rule. Listed below are several types of collateral.

1. Securities of closely owned companies. The problem here is that if the collateral must be sold, there may be no established market available. A particular buyer must be found. Sometimes this obstacle can be overcome if a responsible third party will enter into an agreement with the bank to the effect that he will buy the note and the collateral from the bank should the borrower default.

2. Accounts receivable. These constitute good collateral if they are current, are due from responsible buyers, and do not fall within the limitations of the law or bank policy.

3. Commodities or merchandise. These constitute good collateral under proper circumstances. Ready marketability, margin, time of proposed sale, care during storage, and validity of lien are the particular matters to be discussed with your banker.

4. Machinery and equipment. In recent years an increasing number of banks have engaged in this type of financing. Whether this be for account of the seller or the buyer, whether it be with or without recourse or reserves depends on the individual application. In general, banks engaged in this type of financing feel it necessary that the following conditions be met:

a. Preferably, the machinery or equipment should be new—not used.
b. A reasonable down payment is required (between 25 percent and 33 1/3 percent of cost).
c. The final maturity period might be as short as 12 months or as long as 60 months, depending on type of equipment to be financed.

d. The machinery or equipment should permit a ready sale at a fair value in the used or second-hand market.

e. The estimated profit, or savings plus depreciation, resulting from acquisition should be adequate to repay the loan over the life of the loan.

5. Real estate and buildings. Although a mortgage on real property is the oldest type of pledge known, there are many banks who do not feel that they can mortgage on commercial or industrial property as the sole basis for a business loan. There are many reasons for this point of view. However, they relate, in the main, to banking laws and regulations which govern this type of advance in detail, and in part, to overall considerations of policy.

Bankers at times have been portrayed as people who are willing to loan money only when you do not need it. A banker must put his money to work constructively, safely and profitably. If you approach your relationship with an appreciation of this point of view, you will be doing your part in forming a sound, profitable and enduring association.

Importance of looking ahead. With respect to the outlook for your own enterprise, you should have more information than anyone else. But your plans for the future, if kept to yourself, can be of no help to your banker. While lenders endeavor to keep abreast of developments, they are not so well posted as managers engaged daily in those fields. An exchange of ideas and information between the banker and businessman, with respect to developments affecting the borrower's line, is helpful to both individuals.

What the banker is looking for in the businessman is a sense of balance. The businessman has a right to expect the same quality in his banker. But if the two never trade their ideas on how things look, neither one will have the opportunity to get a "feel" of the other's mind. Without this "feel" it is difficult to achieve that mutual understanding which is essential to a well rounded banking relationship.

The Cash Budget

The cash budget (Table 3.2) can be used effectively when dealing with bankers. Many businessmen do not understand how this is prepared. The one reproduced here is a 120-day cash budget prepared by estimating your assets and liabilities in advance. Past financial records and ratios are very helpful in preparing this budget.

TABLE 3.2.   CASH BUDGET—ABC CORPORATION

Account	March	April	May	June	July
Cash	30,000	30,000	30,000	30,000	30,000
Accounts receivable	60,000	70,000	80,000	90,000	80,000
Inventory	240,000	240,000	270,000	275,000	240,000
Current assets	330,000	340,000	380,000	395,000	350,000
Fixed assets	40,000	40,000	40,000	40,000	40,000
TOTAL ASSETS	(370,000)	(380,000)	(420,000)	(435,000)	(390,000)
Accounts Payable	180,000	180,000	210,000	235,000	200,000
Notes Payable	30,000	30,000	30,000	30,000	30,000
TOTAL LIABILITIES	210,000	210,000	240,000	265,000	230,000
NET WORTH	160,000	160,000	160,000	160,000	200,000
TOTAL LIABILITIES AND NET WORTH	(370,000)	(370,000)	(400,000)	(425,000)	(430,000)
Need to borrow	0	−10,000	−20,000	−10,000	+40,000

Start with the minimum amount of cash you will need. This should be sufficient to cover current commitments, wages and salaries, and a reserve for emergencies. The amount assumed in this case is $30,000.

Accounts receivable should be based on past records for the month and predicted sales forecast. Do not overlook the condition or age of your accounts receivables. If you normally have $25,000 of parts and service sales per month, and receivables are usually two months' parts and service sales, then $50,000 should be the amount of receivables at the end of the month.

Inventory requirements are forecast on the basis of the stock required to service your predicted sales. Again prior years' records are helpful.

The sum of cash, accounts receivable, and inventory will equal your total current assets. Add fixed assets to this to give your total assets. Show this figure in red or in parenthesis.

The next estimate is your accounts payable. This must be based on the volume of business forecast from past records. You should be able to get a ratio between sales and accounts payable. You will normally owe your suppliers a certain percentage of sales at the end of the month. Add your fixed liabilities to your payables to get total liabilities.

Assume your net worth will remain constant over the period of the loan. Add this amount to the total liabilities to get total liabilities and new worth. Show this figure in red or in parenthesis.

In a balance sheet, total assets must equal total liabilities and net worth. If total liabilities and net worth are less than total assets, the difference must be made up by borrowing.

This type of budget will show the banker what you need, when you need it, and when and how you will pay it back. Extend the budget until the balance is such that the loan can be repaid.

Management Practices

Financial management involves many accounts and techniques. As a minimum, remember always to take your cash discounts. Remember that a 2% discount for payment within 10 days is equivalent to 2% for 20 days or an annual interest rate of 36.5%.

Get a daily cash report from your bookkeeper showing cash on hand, accounts receivable, inventory and accounts, and notes payable. This will give you an opportunity to take corrective action and avoid crisis cycles.

Protect your business with insurance from losses that may be crippling. You should consider public liability, fire, and theft as a minumum.

Plan to retain sufficient profits in your business to finance your anticipated growth rate. It takes an increase of 10% in your net worth to handle a 10% increase in business volume.

If you are short of working capital, minimize your fixed assets. Rent, lease, or mortgage them. The use of cash flow budgets will do much to limit impulse buying that deflates working capital.

Finally, visit your banker regularly. Bring your financial statements along. Keep him posted on your operation. He will be in a better position to lend you money if he has a good understanding of your operation.

QUESTIONS

1. Define capital.
2. What is working capital? How can it be increased?
3. Discuss the importance of balance in managing capital.
4. How can you improve your banker's confidence in you?
5. What does the banker look for before making a loan?
6. Why is "cushion" in a loan important to the banker?
7. What are acceptable sources of collateral?
8. How do you prepare a cash budget?

4

Building Location and Requirements

A building and grounds form a very necessary expense in this business. The cost should be no more than that required to fulfill its purpose. Everything about the building and its location should contribute to the purpose, which is to

1. Serve the customer better
2. Aid in coordinating and accomplishing the functions of sales, service, parts, and administration
3. Increase profits

Determining exactly what will enable you to serve the customer better and increase your profits is anything but simple.

LOCATION

The matter of serving the customer better involves many phases of the building, but probably location is paramount. The first question to resolve is whether to select an urban or rural site. The center city site is often as much as 500% more expensive than comparable square footage beyond the city limits. Does the center city location serve your customers better? In some instances it will. The combination of high land values and traffic congestion in cities will almost always, however, cause you to look for a site at or beyond the city limits.

Your customers like to combine trips to town. Look for a location that is well traveled. Most marketing men would advise selecting an area near your competitors and other businesses that may attract your clients. Consider the frequency and driving time of your customers to this location. Consider the ease of maneuvering heavy farm equipment in and out of your location. Finally, consider your own requirements. The convenience of your employees should not be neglected. Neither should the convenience to bulk shipping terminals be overlooked. Mass transit of goods coming in will play an ever-increasing role in the future.

Lot size and topography must be considered. Though the national average is only approximately 60,000 square feet or about 1.5 acres, as dealerships are continually getting larger a tract of up to 5 acres is probably desirable. Three acres should probably be a minimum. A rectangular shape with a frontage of at least 300 ft is ideal. Good drainage on a level to gently sloping lot completes the requirements. It is seldom that the ideal location can be secured.

Additionally, it is essential to consider zoning requirements of the neighborhood. Will your business be compatible with the other businesses in the area? Are you protected against future developments?

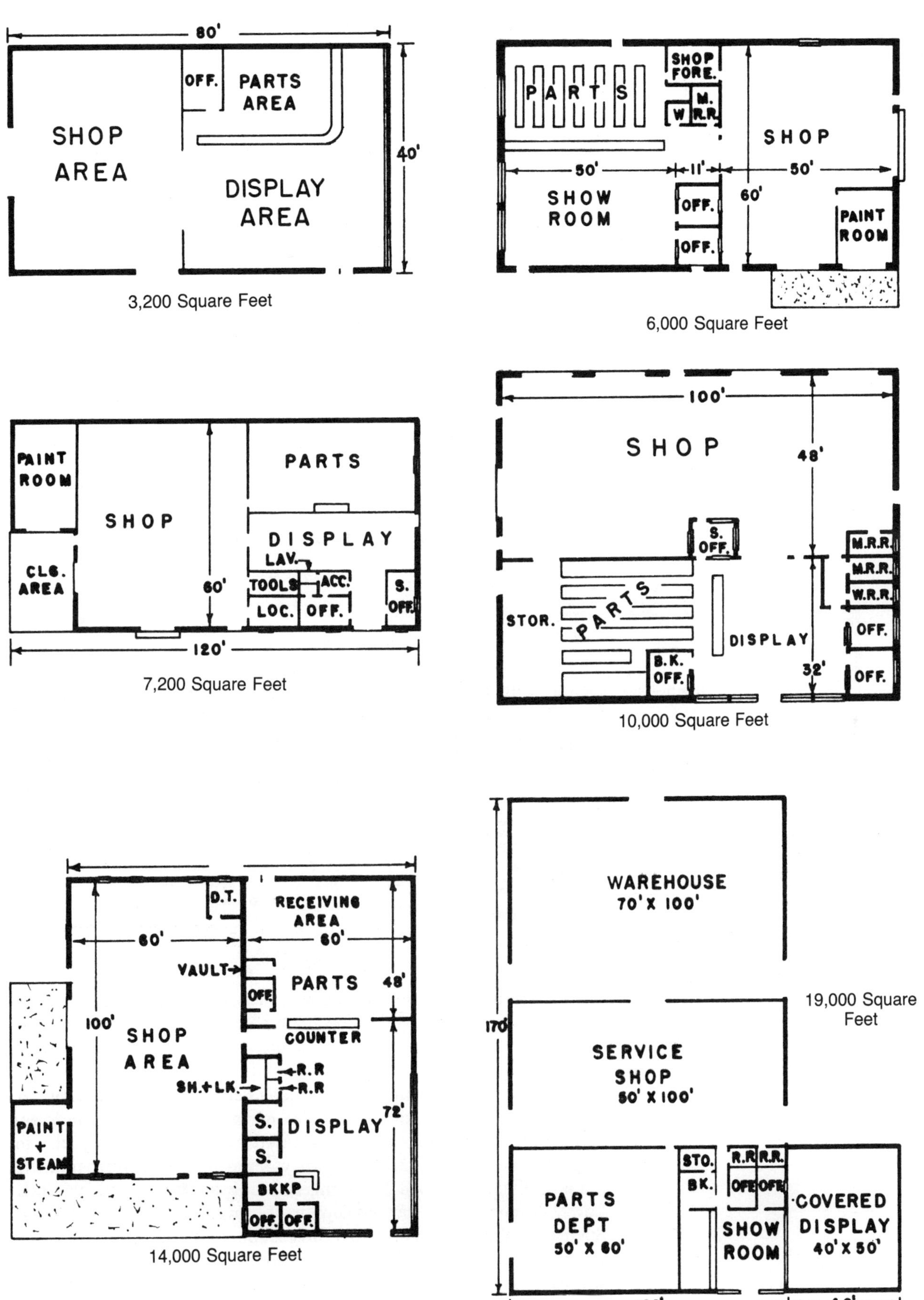

FIG. 4.1. Sample layouts for various sizes of building

What are the long-range plans for roads in the area that could affect your location? By all means, do not overlook utilities. An ample supply of gas, water, and electricity is a must.

BUILDING DESIGN

Four primary zones should ordinarily be included in the main building:

1. Sales area
2. Administrative area
3. Parts area
4. Service area

Additionally, there are numerous secondary areas to consider:

1. Parking
2. Receiving
3. Display
4. Internal road net
5. Storage
6. Identification
7. Sewage disposal area
8. Miscellaneous services such as painting and washing

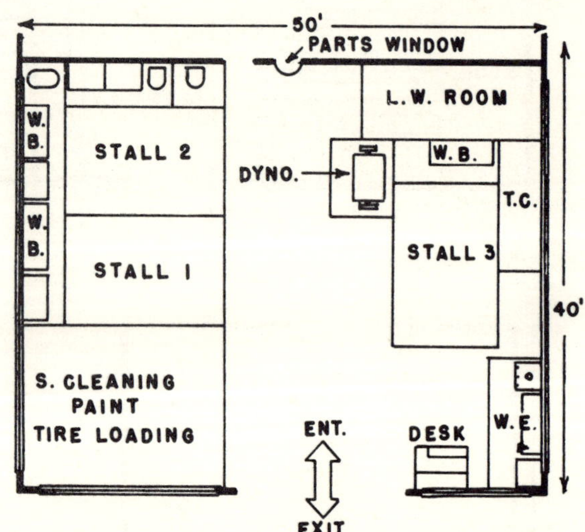

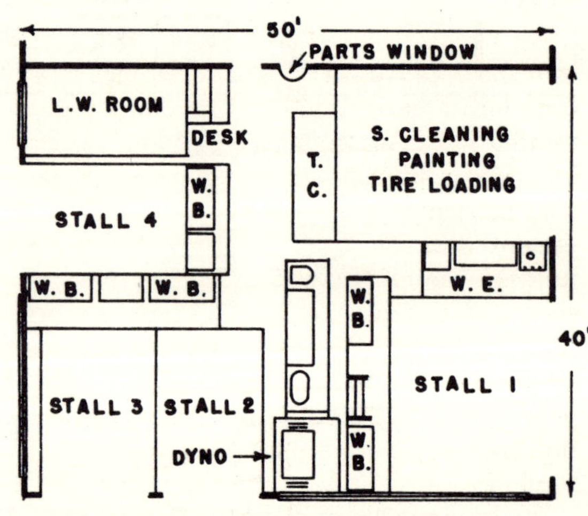

FIG. 4.2. A. A shop suitable for 3 working stalls can be fitted into 2000 square feet. B. The use of outside doors increases effective working space by 25% in this 2000 square foot shop.

The size of the building and auxiliary features is a function of the volume of business. Some examples of various sizes and building layouts are shown in Figs. 4.1–4.3. Although it may be necessary to restrict the size of the building to the volume of today, be sure that the plans include how to enlarge the building for tomorrow's volume. Low-volume dealers operate comfortably with 5000 ft^2 of building. Such a building would probably have a 200 ft^2 shop suitable for three working stalls. A 10,000 ft^2 building would probably have a 4000 ft^2 service shop suitable for seven working stalls. Large dealerships may have 20,000 to 25,000 ft^2 of floor space to work from. There is very little correlation between volume of business and floor space. Ample floor space is certainly conducive to serving customers better and to easing the coordination of activities.

As important as the size of the building are its arrangement and proportions. Figure 4.4 shows the primary and auxiliary functions

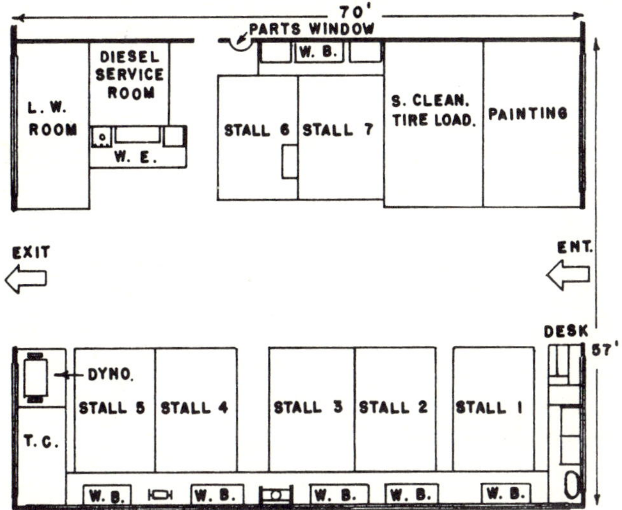

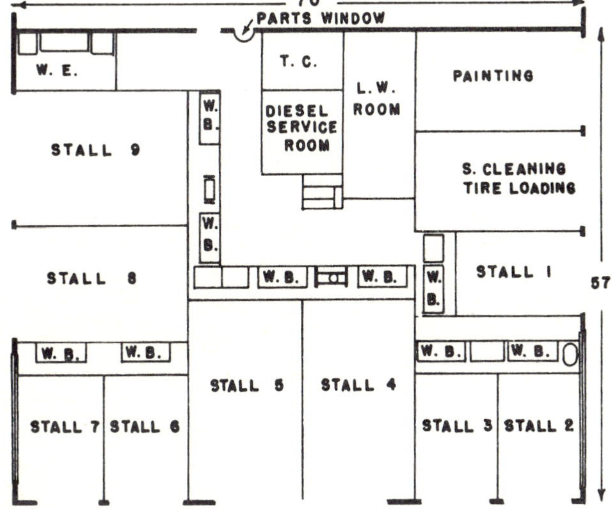

FIG. 4.3. A.   This 4000 square foot shop houses 7 working stalls.

B.   By using outside doors, this 4000 square foot shop can house 9 working stalls.

that the improved portion of the site must serve. The important thing is to locate each functional area adjacent to the area most closely related to it. Administration should be centered, with parts, sales, and service all conveniently nearby. The entry should be such that both sales and parts see the potential customer. A browsing area at the entry for literature as well as featured merchandise and impulse items can entertain the customer until a salesman is free. The service shop needs access to parts. The parts and shop areas should be contiguous, but separated. All areas must have access to washrooms. To the extent feasible, separate dressing room facilities should be provided for the service shop. The coffee break has become an institution in America. A suitable place for a coffee, a lunch room, and a conference room are valuable additions to the structure.

Well-planned stores according to a booklet, "Store Arrangement Principles," published by the Department of Commerce usually have the following characteristics.

1. A suitable and attractive storefront
2. Good signs to identify kind of business (Fig. 4.5)
3. Clean, attractive store windows suitable for the types of lines handled
4. Easily accessible entrance; door at street level
5. Adequate illumination of displays and service shop
6. Department identification to permit easy customer progress
7. Use of light, color, and space to create the impression of size
8. Arrangement of department displays to stimulate sales of related items
9. Proper placement of parts and shop departments to facilitate customer service
10. Accessibility of island and counter displays to invite self-service, if desired

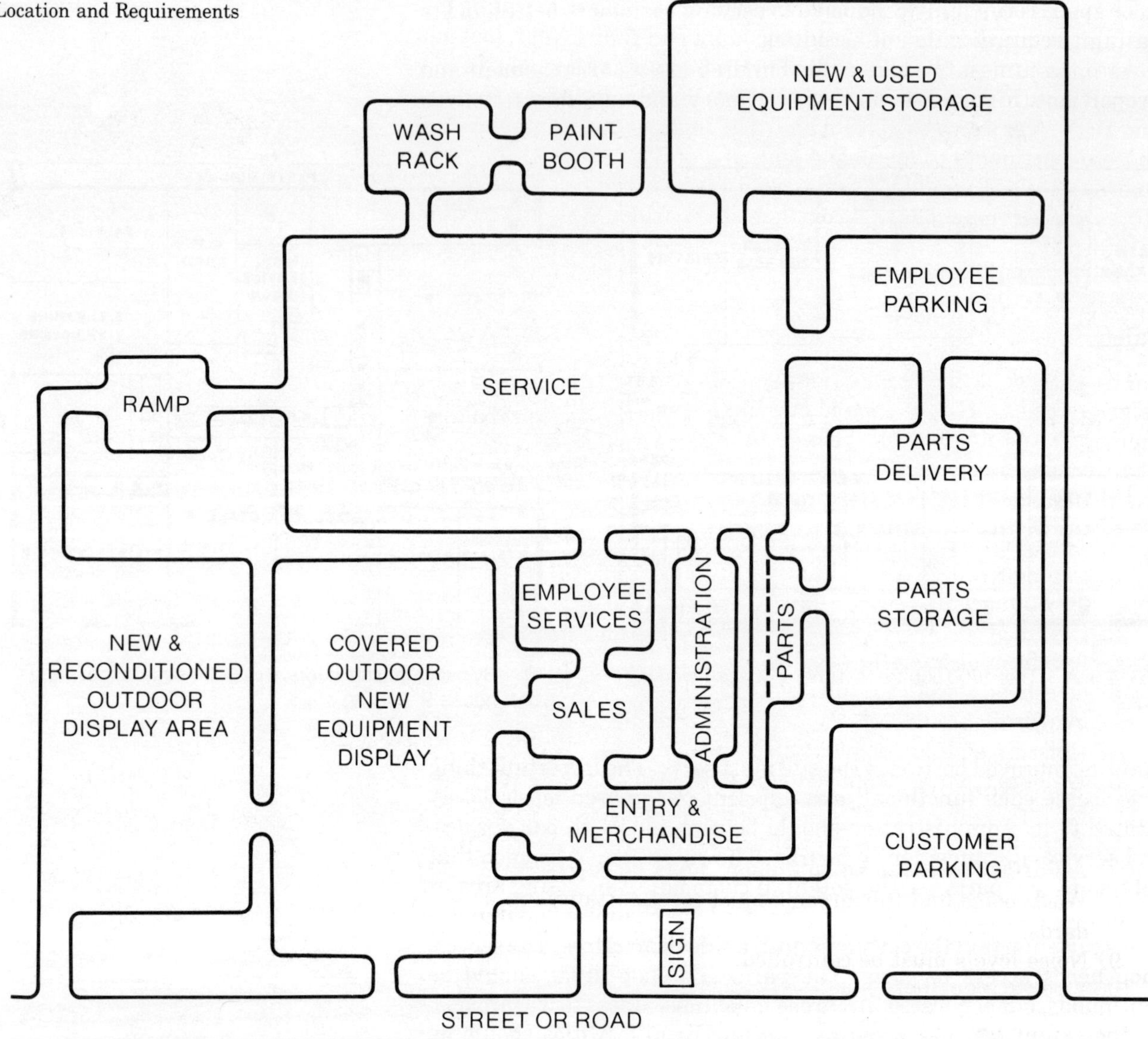

FIG. 4.4. Each area of a well-planned dealership is located adjoining the most closely related phases of the business.

11. Avoidance of excessive visibility of shop operations, such as painting, cleaning, etc.
12. Modern shop facilities to assure customer precision workmanship
13. Prominent displays of parts to create customer confidence
14. Loading platforms and ramps to reduce labor
15. Elimination of hazards to life, limb, or property of customers and employees
16. Judicious use of posters, placards, etc.
17. Customer conveniences, such as telephones, lavatories, and toilets
18. Adequate parking facilities to meet changing conditions
19. Arrangement of departments to avoid unpleasant delays; adjustability to meet peak loads
20. Conformance, in general, to habits of customers

The good retailer blends all of these characteristics in planning his building requirements and designing store and shop layout, looking toward the utmost total advantage to his business as a whole. In the overall picture some weight is given to each item, each department, and the dealer's own skills and abilities. In this way, he emphasizes those departments of merchandising and service in which he feels that he excels. The many blends that result are typical of the many stores and service shops that continue efficiently to serve the American farmer.

Safety

The passage of the Occupational Safety and Health Act in 1970, referred to as OSHA, has added new dimensions to building requirements. The building must be safe, and the environment must be healthy. The following are a few of the features that are most apt to cause trouble with the OSHA inspector.

1. The electrical wiring must meet modern code requirements. A common fault in old buildings is the lack of separate grounding for all outlets.
2. Layout of building must be such that aisles can be kept clear and free from obstacles.
3. Overhead areas and stairways must be properly equipped with guardrails.
4. Hoisting equipment must meet minimum standards.
5. Welders must be shielded to prevent flashes to adjacent workers.
6. Paint booth requirements are stringent and must be met.
7. All drives on installed equipment must be shielded.
8. Washrooms and lunchrooms must meet certain health standards.
9. Noise levels must be controlled.
10. Fire extinguishers must be properly hung.

A careful study of the OSHA regulations will do much to reduce friction when your OSHA inspector comes calling. The "General Industry" standard is available at no cost from your nearest OSHA office.

ESSENTIAL FUNCTIONS

The following are basic ideas regarding the essentials of a good building. Seldom will all be available. The trick is to compromise on the less important ones.

1. Manager's Office—The manager must coordinate the activities of sales, service, parts and administration. His office should be central to these areas. He should be able to see yet he must have privacy.

2. Administration—Like the manager's office, it must be central to the functional units of sales, service, and parts. It should be next to the manager's office. An area big enough to store records and to provide a comfortable working area is needed. It should be pleasant and well lighted.

3. Sales Area—Salesmen should have sales cubicles at the side of the display floor as a minimum. A private sales office with literature

and suitable for showing sales films can be very helpful. It should command easy access to both the display floor and the display lot.

4. Display Floor—The display floor should have space for minimal machine displays. It can probably function best for planned parts and attachments that can create impulse buying. A well-lighted, all-weather outside display is best for large machines.

5. Parts Department—The parts department should be visible from the entrance of the building. It must be convenient to the service shop, but there should be no access to it from the service shop. Steel bins are ideal for parts storage. Some special service shelving is always necessary. A receiving room to inventory and sort incoming parts is a must. Overhead storage is adequate for slow-moving parts.

6. Service Department—A manager's office should be provided with space for service personnel meetings. Privacy is needed when talking to customers. Additionally, facilities must be provided for processing work orders, service manuals, correspondence, and other administrative work expected of the service manager.

The service shop usually occupies half of the main building. Modular construction can make it more efficient. Dwight Horner, AIA, who did a study for Farm and Power Equipment recommends 14- to 16-ft wall heights with 24-ft bay spacing modules and 50-ft clear span construction. Where combines are handled, door heights should be no less than 13 ft with widths to accommodate the largest headers. The 24-ft module is ideal for two tractor stalls and will accommodate most combines. The length of the shop can be as long as needed to handle the service volume. The constantly increasing equipment size may dictate going to a 60-ft clear span. This allows room for a center aisle

FIG. 4.5. A prominent identification sign enhances the business.

for foot traffic. Exterior doors prevent traffic jams. Vary the size of the doors in each bay to accommodate individual needs.

Pay particular attention to the electrical wiring. A circuit breaker type main service panel with room for expansion is needed. Use underground wiring for service to outdoor display area, spot lights, neon signs, and yard lights. Provide 220-V service for air conditioning, welding machines and other special tools. The paint and pressure-cleaning areas require considerable power. Convenience outlets located overhead over the work stalls or around the periphery of the building are needed. Light panels overhead and overhead lighting are needed to provide proper lighting for efficient work.

A "clean room," air-conditioned for service on diesel pumps and injectors and hydraulic equipment; a tool room for specialized tools; a painting and cleaning area; and a locker room and rest room for employees complete the main parts of the service area.

7. Lunch Room—Today employees expect to be furnished a place to relax during coffee breaks. The lunch room is a morale-building factor that should not be overlooked. It has the added advantage of providing an excellent place for after-hours training sessions with employees.

8. Storage Room—The storage and set-up room can be provided separately from the main building or an integral part of it. It need not be as expensive construction as the main building. Security of contents, good working conditions for set-up work, and convenience for bulk handling of heavy parts with fork lifts and hoists are most important.

A building is a personal thing. The use of a professional to design your building is usually a good investment. It is necessary that you give him some guidance as to your desires.

Conclusions

A building and grounds must serve the customer better and help you to coordinate and accomplish the functional activities of the business and to increase the profitability of the business. Time spent studying existing designs and visiting other dealers can help tremendously. Remember your supplier is also a good source for ideas on building needs. When you are reasonably sure what you want, get an engineer or architect who has experience in small commercial buildings to prepare plans and specifications. Professional help really pays off.

QUESTIONS

1. What is the purpose of the building in a retail dealership?
2. Discuss the merits of "close in" versus "outskirts" locations of dealerships.
3. What primary zones must be included in a building?
4. Discuss the merits of separate parking lots for customers and store personnel.
5. What part does OSHA play in the operation of a sales store?
6. Discuss the need for a service supervisor's office.

Selecting Lines of Merchandise

Dealerships usually stock one long line and one or more short lines of equipment. A manufacturer who produces both power units and a complement of implements to produce and harvest crops is said to offer a long line. The manufacturer who specializes in only a limited type of equipment is referred to as a short-line manufacturer. Both are very important. The proper selection of long and short lines is vital to the success of a business. The key is finding lines that will have utmost local acceptance. National advertising along with proven performance facilitates getting local acceptance.

The discussion always crops up as to which is most important—the dealer or the line. The question is really academic. Both are most important. Certainly a line must be acceptable in a community. This is unknown until someone makes the determination. Since this is costly, many would prefer to pay extra to buy into a franchise of a brand that has demonstrated local acceptance. The desirability of having a proven product cannot be questioned. It certainly is obvious that a dealer with a proven ability to sell and service farm equipment is in a better position to launch a new brand than an unknown and unproven dealer. Both the brand and dealer must become a team if a dealership is to be successful.

The business is built around the personality and performance of the dealer. He must be sold on the line he represents. The customer must be sold on the reliability, the integrity and the performance of the dealership.

The business needs a broad line. The tractor, upon which most of the image of the dealership is based, accounts for no more than 40% of the sales volume. It is the many other implements and machinery handled that make possible a profitable volume of business. Diversification is not a panacea, but it does offer several advantages.

1. It increases store traffic. Customers get in the habit of coming to the diversified dealer, where they know all their needs will be filled. Impulse buying usually results when equipment is well displayed.
2. A larger volume spreads the cost of doing business over a broader area. This makes possible the hiring of more efficient help and its full utilization over a longer period of time.
3. It levels seasonal humps. Planned diversification can balance labor and capital needs over the entire year. Improperly planned, it could have exactly the opposite effect.

RESPONSIBILITIES TO A LINE

The primary responsibility of a dealership is to support the "long line" handled. Any additional goods or service handled imposes additional responsibilities, which the organization must be prepared to support. The following are the major considerations.

1. Requires extra capital. Capital, like other tangible assets, cannot be used two places at the same time. Extra lines require an inventory of goods, parts, and personnel. This translates to extra capital. The amount needed will vary with the equipment, but it's a safe assumption that the capital requirements will not be very different from the rest of the business. This translates to from $12,000 to $14,000 for each additional $100,000 of annual sales expected per year, or $1.00 of capital for each $7.00 to $8.00 of sales.

2. Need personnel. Unless you have employees who are currently doing nothing, extra help will be required. Some seasonal items may be dovetailed into the business to minimize this. A careful analysis should be made before assuming that the current staff can handle the additional business.

3. Need specialized training. Salesmen can sell and servicemen can service, but only after they become completely familiar with the new line. This requires specialized training. Don't damage your image by doing less than the good job you are doing with your main line.

4. Need additional customer financing. The new line sounds good to you, but is it good enough that customers can find ready financing for their needs? Don't overlook this important aspect of retailing.

Along with the responsibilities, a few words of caution are in order. Don't water down your entire effort. Set up the other line as a separate department. Put someone in charge and let it carry its own weight. If it's profitable, you will soon know. If it isn't, remember the good businessman cuts his losses short. Diversification can work wonders, but only if it is based on community needs and the dealer's own needs.

When the decision to add an additional line has been made by management, an all-out effort must be made to support it. This encompasses most of the things required to support the main line. Let's summarize.

1. Learn all you can about your new product and develop a sales campaign for it. It's good and it's needed but you still have to sell it.

2. Give your employees a thorough orientation on the new product. They too have to be sold on the new line. You can't sell yourself on anything until you know it and understand it.

3. Stock a balanced inventory of goods and parts. Customers like to see what they are getting. They want to know that you plan to stay in the business. What better way to show that than by stocking a good selection and ample parts for servicing.

4. Display it, advertise it, and demonstrate it. How else will people know that it is available and will perform its intended function? Your main sales advantage over the mail-order catalog is that your merchandise can be seen performing its job before it is bought.

5. Arrange financing. Who are your customers? Can they afford to pay cash for their equipment needs? If you are selling only to cash customers, you are missing probably 90% of the potential sales. Go to the bank or other lending institutions and work out a financing plan for your customers. Let your customers know in your advertisements that you can finance their needs.

6. Maintain a good parts inventory. Check your distributor for parts. After all, you can only afford to carry parts that move at least twice a year. Will he back you up with quick shipments? Is the parts pipeline open all the way to the manufacturer?

7. Maintain a good service shop. See that your mechanics are factory trained on all new sophisticated items. The success of a new line may well depend on the performance of the men in the shop.

8. Departmentalize your accounting. Good management is possible only with accurate timely figures. Departmentalized accounting will allow you to manage the new line.

FIG. 5.1. Diversifying into other lines such as grains bins may help even out the work load.

DECIDING ON A NEW LINE (FIG. 5.1)

Selecting the new line requires careful analysis. Not only must you decide what line but what make within the line. Let's consider some of the factors that would influence your decision.

1. Select a line suited to the area. Verify before stocking that there are customers. Convince yourself that the line will economically benefit the owner either in money or additional leisure time.

2. Select a line related to your current line. This will minimize training requirements as well as specialized shop and display facilities.

3. Select a line that is not currently adequately represented in the area. It is not good business to stock equipment that is in oversupply. However, the fact that there is competition is no reason to avoid a line. The key term is adequately represented.

4. Select a line that is nationally advertised. Surveys conducted to determine where people first hear about new equipment show that 65% hear about it first in farm magazines. The importance of national advertising is shown to be very cogent when we consider that the same survey indicates machinery company literature only accounts for 48%. The local dealer ranks highest (51%) in presenting the most useful information, followed by the farm magazines with 42%.

5. Select a line that has a future. Don't wait until the local demand has been satisfied before entering the market. Cropping patterns, crops and techniques of handling are in a constant state of change. Keep current on agricultural affairs. Your county agriculture agents and the various USDA men working in the county can be most helpful in this field.

6. Select a line that the manufacturer backs. Warranty terms are a good starting point. Additional supply of parts and adequate inventory of whole goods is most important. Does the manufacturer help with local advertising? Is training available for service men? Is ample literature available for potential customers? Floor plan terms and customer financing services are other important considerations.

7. Select a line you believe in. The salesman has to sell himself before he can sell anyone else. Sincerity in the sales presentation can often be detected readily by the customer.

8. Select a line that can be financed. Most of your customers need financing. Are there financial institutions that will lend money on this line?

9. Select a line you can support. Support is a broad term. Remember, you will need about $1.00 of working capital for every $7.00 of sales. Parts must be binned, stock must be housed, repairs require shop facilities and mechanics, and salesmen have to be trained before a line can be successful in an area.

ADDITIONAL LINES

Other lines are as plentiful as there are innovative and perceptive dealers. Choosing another line is an individual thing. Many dealers have found items on the following list profitable, although the list is far from complete. It is shown mostly to promote thought on the subject.

1. Light industrial equipment
2. Lawn and garden tractors
3. Hardware
4. Feed, seed, and fertilizer
5. Crop-drying equipment
6. Grain storage bins
7. Irrigation equipment
8. Pesticides
9. Dairy equipment and supplies
10. Farm fuels
11. Chain saws and timber harvesting

QUESTIONS

1. Differentiate between "short-line" and "long-line" equipment.
2. What responsibilities does a dealer have to a line he stocks?
3. About how much capital is required to handle a line?
4. What is necessary in a dealership in order to properly support a line?
5. How would you determine which brand of merchandise you would choose to stock?
6. How would you determine if a line would augment or compete against your main line of merchandise?

6

Adequate Insurance

The world we live in makes insurance mandatory. It is unfortunately too true that our ability to pay insurance premiums is often inversely proportional to our insurance needs. Nonetheless, it is a business expense that must be borne by the struggling businessman even when the premiums reduce badly needed working capital.

PLANNING COVERAGE

For a fee, any peril can be insured against. Every dealer certainly does not need this level. A program tailored to his needs must be worked out. It is a specialized and technical field. Have a competent insurance agent survey all the risks and make recommendations. In the final analysis, it is the manager's job to determine what insurance to buy.

Risks may be classified as follows:

1. Financially crippling losses
2. Seriously handicapping losses
3. Moderate losses
4. Nuisance losses

It is evident that the first two perils should always be guarded against. The latter two may or may not be, depending on the situation. Only the individual dealer taking into consideration his own resources can determine what perils would seriously affect his ability to carry on his business. What may be a nuisance to a well-established and well-financed dealer may be a seriously handicapping loss to a struggling young dealer just getting started.

Averages are deceiving, but for lack of a better guide it may be informative to look at the magnitude of this expense to the average American dealer. According to the 1980 Cost of Doing Business Study, nongroup insurance represented an expenditure equal to 0.51% of sales. Group insurance probably adds another 0.25% of sales. There were differences when insurance was considered as a percentage of sales according to volume of business conducted. The percentage tended to go down with increase in volume. It is interesting to note that the high-profit dealer carried less insurance than his

TABLE 6.1. COST OF INSURANCE AS A PERCENTAGE OF SALES (1980) (NON-GROUP)

Category	Average	High profit	Low profit
All dealers	0.51	0.48	0.67
Up to $1,000,000	0.73	0.52	0.91
$1,000,000 to $2,000,000	0.61	0.62	0.64
$2,000,000 to $3,000,000	0.50	0.48	0.46
Over $3,000,000	0.43	0.41	0.49

low-profit counterpart, both in dollar value and as a percentage of sales.

An investment of this magnitude requires careful thought and management. Risk must be carried by the dealer or insured against. In order to decide what insurance is best for you, you should know all the insurable business risks of your operation, how to insure against them, and the cost of insurance. You must choose a competent insurance broker or agent as an advisor. Choose only one advisor. He can advise you as to what insurance will protect you against the various perils and what it will cost.

Your first step with the help of your advisor is to make an insurance survey. He will complete a study of your business, getting you to provide him with the information he needs to discover your insurable business risk. This is the normal service expected of an insurance agent. There is no charge for it.

An insurance survey should provide the following information:

Important insurable risks you might be leaving uncovered
Unimportant risks you are covering that should be dropped
Values that are out of date and in need of revision
Duplication of coverage among different policies
Errors in your policies—rates, names, addresses, and so on
Whether you are buying your insurance at the most economical
    rates
Whether your rates could be reduced by modifying your methods of
    operation

Insurance is a complicated matter. General discussions of insurance needs serve a useful purpose. They do not, however, reduce the need for competent on-the-spot advice based on your particular conditions.

Insurance may be divided into three general categories, liability, property, and life and health.

All of your liability insurance should be handled by one company. It is not necessary to place all coverage with your advisor's company. Property damage may well be placed with some other business associate. In any event, let one man be responsible for advising you about your needs. This will usually result in maximum coverage at a minimum cost.

LIABILITY INSURANCE

As a farm and power equipment dealer, you have six major liability exposures, which grow out of these risks:

Your use and maintenance of your premises
Your operations involved in conducting your business both on and
    off premises
Your use of owned and non-owned autos, trucks and other equip-
    ment
Your liability for products you sell
Contractual liability
Liability for the safety of your employees

The Garage Owners policy is suited to the retail farm equipment dealer. It is designed to cover his legal liability for accidents arising from his operations and the use and maintenance of his premises. Operations away from the premises are covered when the operations are necessary or incidental to the business described in the policy. Legal liability for accidents resulting in connection with mechanical repairs to equipment or parts are covered also.

Vehicles probably constitute the greatest risks the dealer is faced with. This policy covers the ownership, maintenance, and use of vehicles used in the business and for occasional nonbusiness use, provided they are primarily used in the business. Basic limits should be set high enough to protect fully the assets of the individual or company.

The Garage Owners policy may be written separately or as an endorsement to a General Comprehensive Liability policy. The latter is preferable since the comprehensive policy provides coverage for unknown hazards or hazards that may be subsequently acquired. It may also include contractual liability. This is quite often involved when dealers accept certain equipment from manufacturers for the purpose of holding demonstrations. If a "hold harmless" agreement is reached with the supplier, then the dealer accepts contractual liability for the equipment. This is not covered under the Garage Owners policy or the Comprehensive Liability unless the policy is endorsed to that effect, making the supplier an "additional insured" in respect to the consigned equipment.

When designing the Garage Owners Liability policy, remember to have it endorsed to cover medical payments. Choose a deductible provision that you feel comfortable with. The higher the deductible, the lower the premium. If there is a grease hoist, you may want to have the policy endorsed to provide collision coverage for damage to property of others in your charge. Pay particular attention to equipment you rent to others. Generally, it can be covered by endorsement under a general comprehensive policy, but you may want to use a separate policy for this risk. Remember that individually owned cars of partners or officers of a corporation should be separately insured.

It has been said that "The big print giveth and the small print taketh away." Certainly it is necessary to read your policy carefully. Pay particular attention to exclusions and endorsements.

The products and equipment you sell go into the hands of the public. You may sell equipment that is not up to standard from a safety standpoint. Rightly or wrongly, the customer may choose to hold you responsible for any accidents that happen while using this equipment. Delicate legal problems are involved here. The courts have held that the manufacturer is responsible for the design of the equipment. The dealer may not knowingly sell unsafe equipment. He has the responsibility to educate the customer in its proper use. Defense of accident claims are expensive, requiring good legal talent and often expensive technical witnesses.

The Garage Owners policy provides product coverage for work performed and material sold in the operation of the implement business only. If you handle other products, you normally would have to cover those with a Comprehensive Liability policy.

There are two very important things to remember about a liability policy. First, liability by legal definition protects you against legal action that may be instituted against you by others. It does not pay you for personal injuries or injuries to your employees. This is available as a separate policy. Second, all of your liability insurance should be placed with one company. This avoids the question of where one company's liability ends and the other's starts.

Injury to employees should be covered with a Workmen's Compensation policy. This is a statutory requirement that varies from state to state. Have your insurance advisor brief you thoroughly on its provisions in your state. Large claims can be generated from this liability. Whether or not it is required by law, you are still liable for injuries sustained by your employees.

Workmen's Compensation is a means of compensating employees and their dependents for accidental injuries or occupational disease received in the course of their employment. If death results from injuries, death benefits are also payable. Benefits are provided for employees through different methods according to the location:

1. Some states provide a state fund whereby the employer pays a premium into the fund based on the payroll of the employer.

2. Some states allow the employers to carry Workmen's Compensation through insurance companies. The premiums are paid directly to the insurance company based on the payroll.

3. Some states allow employers to be self-insured if approved by the proper state authorities.

4. Many states use a combination of these methods. In some states, benefits under Workmen's Compensation are rather modest. Limits of $25,000 for death benefits are not uncommon. Prudent dealers are buying separate liability coverage. If an employee suffers a severe injury or has a basis for a civil suit, he may choose to sue for an amount exceeding the limit. Protect yourself against such a possibility by increasing your Employer's Liability limit. The cost of higher limits is very nominal.

PROPERTY INSURANCE

Property insurance is designed to reimburse you for loss or damage to property by fire, explosion, windstorm, hail, glass breakage, theft, and similar hazards. The amount of insurance to buy and which hazards to protect against should be determined in consultation with your insurance advisor. Certainly, the extent of your investment in plant, buildings, and equipment must be considered.

Property values due to inflation have increased from year to year. Review your coverage periodically to make sure you are amply covered. Of equal importance is the problem of double coverage. Make certain that equipment that is floor-planned is covered by the supplier or by you but not by both.

Property insurance policies are usually purchased as a Fire Insurance Policy. This coverage can be extended by endorsement to cover property loss from windstorm and hail, explosion, riot, strike, civil commotion, falling aircraft, vehicles, smoke, and other causes.

An endorsement is also available for Vandalism and Malicious Mischief, as well as for Theft and Glass Breakage.

Your policy should specify the property it is to cover, such as buildings, equipment, machinery, and stock. Various companies handle this in different manners. Usually values of stock are reported on a monthly basis with the premium assessed accordingly.

Insurance rates are based on the experience of a company in an area. High losses result in high premiums. Property insurance rates are based on type of construction, fire protection available, wiring, and the like. A good insurance advisor can show you how you might add some protective devices and reduce you premiums. In the long run, this results in lowered operating cost.

How Co-Insurance Works

Many property insurance policies are written subject to a co-insurance clause. This is an agreement between the dealer and the insurance company to the effect that in consideration of the buyer maintaining insurance to at least a certain percentage of the insurable value, he (the buyer) gets reduced premiums. If the buyer fails to maintain the minimum value, he is penalized on a partial loss. In the case of total loss there is no penalty. To get the benefit of the lower rate, he must carry the agreed amount of insurance.

Consider the following example for a policy with an 80% co-insurance clause:

Insurance value of building	$50,000
Insurance carried	40,000
Loss	10,000
Amount of loss paid	10,000

Consider the example when the insured amount is allowed to drop:

Insurance value of building	$50,000
Insurance carried	30,000
(with 80%) co-insurance clause)	
Loss	10,000
Amount collected	7,500

The insured is carrying $30,000 of insurance and should be carrying at least $40,000. He is three-quarters insured (30,000/40,000). He collected only $0.75 \times \$10,000$ or $7,500. It is very important, particularly in times of rising values, to check to see that you are within your co-insurance requirements.

Co-insurance is a form of quantity discount. Insurance companies know that most claims are usually for less then the total value of the property. Where good fire protection facilities are available, a co-insurance clause can reduce premiums as much as 20%.

Care should be exercised where rented or leased buildings are involved. A dealer may be held liable to his landlord for damage to the building if it can be shown that the fire was the result of negligence on the part of the dealer or his employees. Protection to the dealer can take two forms:

1. A waiver in the lease agreement
2. Purchase of "Fire Damage Legal Liability" insurance. This may be an endorsement to a Comprehensive General Insurance policy

The loss of a building usually results in suspended business operations. Consider Business Interruption insurance. This insures you against loss of earnings for the length of time a business is partially or totally suspended by one of the perils insured against. The amount of insurance written is based on the dealer's estimated annual earnings. This policy may have a co-insurance clause of from 50 to 80%.

Business interruption insurance may take the form of earnings insurance. Policy limits are set by estimating the highest monthly earnings for the coming year and multiplying this by the number of months needed to get the business back in operation. Normally a dealer cannot recover more than 25% of the policy limits in one month. This may be applied to payroll, expenses, or profits.

An Installment Sales Policy is available to reimburse a dealer for loss or damage to equipment and merchandise sold at retail on an installment contract. Coverage should be written to cover the interests of the dealer and the customer. The certificate type of policy is issued to a dealer, who in turn issues a certificate of insurance to his customers covering their purchase on installment or conditional sales contract. This gives the purchaser evidence of the insurance applying to his purchase. Coverage is normally on an "all risk" basis subject to standard exclusions. Since the customer is responsible for items he buys, any loss sustained is legally borne by the purchaser and not the seller. It is nice to have coverage that promotes good public relations but not as necessary as some other coverage.

ACCIDENT AND HEALTH INSURANCE

Good personnel relations can be enhanced with good group Accident and Health and Life Insurance. A program of this type has come to be expected by employees. Certainly there are several good reasons for putting on such a program.

1. It relieves the employee of worries, improving his efficiency as a worker.
2. It attracts and helps keep good employees.
3. It removes the dealer from a charity role in relation with employees.
4. It promotes good will.
5. It lowers the cost of insurance to the individuals.

The National Farm and Power Equipment Dealers Association sponsors accident and health insurance in cooperation with state associations. Their large purchasing power brings the cost down to modest levels. The high cost of hospital and medical care makes even this minimum rate a rather substantial item.

Accident and Health Insurance pays for expenses in connection with sickness or accidents as well as providing an income during sickness or injury. The most common forms of coverage sold are accidental, health and dismemberment, hospital, surgical, loss of time, dread disease, x-ray and laboratory, and medical expense benefits. Major medical coverage superimposed on the basic policy is also available to cover more costly illnesses. Since employees are covered under Workmen's Compensation, an exclusion is written into the policy to cover accidents on the job. This reduces the cost somewhat.

According to the Farm and Power Equipment Dealers Association, the following coverage is available as a package policy to all farm equipment dealers:

Life Insurance
    Owners
    General Managers
    Eligible employees
Accidental Death and Dismemberment
    For each enrolling

Sickness and Accident Benefits
  Stated amount per week
Hospitalization Insurance
  Semi-private room rates paid in full
  Reasonable and customary hospital charges
  Employees and dependents
Surgical Operation Benefits
  Reasonable and customary charges for surgery
  Employees and dependents
X-ray and Laboratory Benefits
  Stated amount for service in doctor's office or hospital
  Employees and dependents
Medical Expense Benefits
  Reasonable and customary charges for doctor's visit
  Employees and dependents
Major Medical Expense Benefits
  Unlimited expenses in excess of those covered under basic policy.
  Benefits are not scheduled but are subject to a deductible and
  co-insurance by the insured.
  Employees and dependents
  Life insurance

There is no scarcity of life insurance policies available to the dealer. The nature of business prompts many to cover their lives and the lives of their key employees as a hedge against loss of earnings caused by their death. These may take the form of individual coverage, business or partnership coverage, group coverage, and coverage of pension plans for employees.

In summary, remember to cover all financially crippling losses as a minimum. In addition statutory insurance requirements such as Workmen's Compensation must be covered. Get a competent advisor and work out your coverage. Keep your values updated. Avoid duplication.

QUESTIONS

1. Why should all your liability insurance be handled by one company?
2. What is liability?
3. What would you include in an insurance survey?
4. What do we mean by co-insurance?
5. What is Workmen's Compensation? What are its provisions in your state?
6. Why should you provide accidents and health insurance to your employees?
7. What does a Comprehensive General Liability Policy cover?

7

Some Tax and Legal Aspects of Retailing

This chapter is not intended to serve as a source of highly accurate data on the subject. It is intended to describe in a general way some of the applicable statutes and to suggest areas that demand very careful examination by your lawyer and tax consultant. Federal laws and regulations affect businessmen the country over. State laws vary considerably from state to state. Because of this situation, the legal and tax aspects of retailing must be approached in a general way.

TYPES OF OWNERSHIP

Retailers generally do business organized in one of four ways—single ownership, partnership, corporation, and cooperative. Each has certain advantages and disadvantages that must be weighed prior to reaching a decision.

Single ownership gives complete control to the owner. That is certainly a nice feature. It also gives complete liability to the owner, which could be bad. The single owner jeopardizes all his assets, both business and personal. Single ownership is taxed only once. This is often an advantage. A small business may choose to take advantage of a corporate structure and still elect to be taxed as an individual. The life of the business coincides with the life of the owner. The will or other succession document is necessary to control the operation of the company upon the death of the owner.

The partnership involves two or more individual owners of a business. It has the advantage over individual ownership in that assets and skills are increased. The partnership is dissolved in case of the death of a partner. This may cause problems due to the sudden withdrawal of a portion of the capital in the company. The problem may be solved by each partner insuring the other for an amount equal to his financial share of the business. Upon the death of one, the survivor has sufficient money to buy out the share of the deceased. The cost is normally considered a business deduction. It is necessary to advertise a notice of dissolution in compliance with the Uniform Partnership Act upon the death of a partner. In the case of partnerships, all of the earnings are taxed to the partners, whether the profits are actually paid to the partners or not. The salaries of partners become income and are not deductible items of expense to the business. The fact that partnership earnings cannot be retained in the business without paying the tax annually makes the corporation form of business more appealing to many. Additionally, each partner may be held responsible for all of the liability arising out of the business.

The corporation, from a legal point of view, appears to have the most favored position. The corporation is a separate legal entity distinct from the owners. The owners are stockholders who may or may not manage the business. Their liability for the acts of the corporation are limited to the value of the shares they own. Personal liability is therefore kept to a minimum. The corporation further enjoys perpetual life since upon the death of a stockholder the shares are merely passed on to the heirs.

The control of the corporation is in the hands of a Board of Directors. The actual operation of the business is done by officers appointed by the board. The corporation must hold at least one annual meeting of stockholders per year, at which time a Board of Directors is elected. At this annual meeting, minutes must be kept.

The capital available to a corporation is greater than for the other forms since shares can be sold to the public. This is predicated on the basis that the corporation has a profit potential and the public is willing to invest in the corporation. Stockholders have no responsibility toward the management of the business.

The details of incorporation vary from state to state. Basically the following steps should be followed. First a name has to be selected and cleared with the Secretary of State or other appropriate state agency. Normally there is a minimum number of persons or "incorporators" who make the application. Each must own stock. A minimum amount of paid-in stock may be stipulated. A typical small corporation may well consist of a man, his wife, and his lawyer or bookkeeper. Some states may require fewer than three. The details of incorporation are more complicated than indicated here. Money spent for good legal advice will be money well spent.

To summarize, we see that the corporation offers the following advantages:

Limited liability of stockholders
Separate legal existence apart from the stockholders
Perpetual life
Possible tax advantages
Ease of transferring interest in ownership

Corporate Taxes

Corporate returns are subject to dual taxation—corporate taxes and then individual income tax on dividends to stockholders. Certain corporations may elect to be taxed as partnerships under certain conditions. The Technical Amendment Act of 1958 added Subchapter S, which permitted small corporations to make this election. The terms of the act require that

1. The corporation can have only one class of stock.
2. The corporation cannot have more than 10 shareholders. This was changed to 25 under certain circumstances by the Economic Recovery Act of 1981.
3. All shareholders must be individuals.
4. The corporation cannot have a shareholder who is a non-resident alien.
5. Not more than 20% of the corporation's gross receipts can be derived from dividends, rents, royalties, interest, annuities, and gains from sale or exchange of stocks or securities.

6. Not more than 80% of the corporation's receipts can be derived from sources outside the United States.
7. The corporation must have the consent of all stockholders.

For equipment dealers who can qualify there is no readily apparent taxation disadvantage, since double taxation is eliminated. Subchapter S is a tax relief measure for small businessmen. Individual stockholders whose individual tax rate is higher than the corporate rate may not find it to their advantage, however.

Corporations are currently taxed at the rate of 16% for the first $25,000 of net earnings and graduated upwards to 46% for all earnings over $100,000. The surtax is eliminated. The tax rate for individuals ranges from 14% for the first $2300 to 50% for all earnings over a certain amount. Tax laws and rates are constantly being evaluated and changed. Keep current on these changes if you wish to minimize your tax bill.

Family Considerations

Estate taxes should be considered in financial planning. Consult competent estate planners for current rules. Tax rates go up sharply as the size of the estate increases. Tax-free gifts can be made by individuals. There is a limit on the amount per individual per year, the total gift, and the time prior to death of the gift. The need for capital in a business often rules out gifts of cash to children during the working lifetime of an individual. In the case of an incorporated business, gifts of shares may be given. In this way the estate taxes can be minimized while the capital remains in the business for use by the business.

Though there are some states and some conditions in which a will may not serve a useful purpose, it is highly advisable to execute a will. Consult a lawyer so that your estate will be divided in accordance with your wishes.

Self-Employed Retirement Plans (Keogh)

Self-employed persons can now contribute 15% of their earnings up to $15,000 per year of earned income to a retirement plan. This is provided for by the self-employed Individuals Tax Retirement Act of 1962 as amended in 1974 and 1981. Funds set aside under this act as well as accrued interest are not taxed to the individual until they are withdrawn. This is normally after retirement when earnings are less. This act is especially appealing to those in the higher brackets. As an example, a taxpayer in the 50% bracket can invest the entire $15,000 instead of the $7500 he would have left after taxes. In addition, he would reinvest the interest on the entire $15,000 tax free instead of receiving interest on $7500 and paying taxes on that amount. Currently many financial institutions are qualified to handle these accounts. The owner has considerable latitude on how the money is to be invested.

INCOME TAXES

Income taxes give rise to certain terms that should be understood. They are listed below for the benefit of the uninitiated.

1. Accrual basis. This refers to the accounting method in which items of income are reported in the year when earned, even

though they may not yet be received; costs are deducted when incurred, although payment may actually be made in a later year.

2. Cash basis. This is the accounting method in which items of income are reported in the year they are actually received and costs are deducted in the year when actually paid.

3. Total income. The general term "income," in the most broad sense, means receipt of anything except return of capital.

4. Gross income. This is total income less certain items that Congress has specifically excluded from tax such as a pure gift, exempt interest part of capital gains, and so on.

5. Adjusted gross income. This term generally applies only to individuals. It means gross income less expenses incurred in the production of income, such as a proprietor's business expenses, business expenses of an outside salesman, or expenses related to investment income.

6. Taxable income. This is the amount of income to which the tax rate is applied. For corporations, it is gross income less costs and expenses. For individuals, it is adjusted gross income less personal deductions and personal exemptions.

7. Exempt income. Congress has the power to tax all income; however, it has specifically excluded certain income from being subject to tax, such as gifts, life insurance proceeds, and interest on certain bonds.

8. Ordinary income. Profit from the normal operation of your business, salary, dividends, interest, and rent received are the principal sources of ordinary income. The ordinary income tax rate for individuals is graduated from 14% to 70%. The rate for a corporation varies from 17% for the first $25,000 to 46% for all over $100,000 (effective January 1979).

9. Capital gain. This is the gain made on the sale of a capital asset or on any property you held as an investment or used in your business, but did not hold for sale to customers in the ordinary course of business.

10. Long-term capital gain. The capital gain made on the sale of property you have owned for more than six months is long term. The long-term capital gain tax-free rate for corporations is 25%; the tax for individuals is computed by taking 40% of the capital gain and taxing it as ordinary income.

11. Short-term capital gain. The capital gain made on the sale of property you have owned for six months or less is short term. It is taxed the same as ordinary income.

12. Capital loss. This is a loss on the sale of a capital asset or on any property you held as an investment or used in your business, but did not hold for sale to customers in the ordinary course of business. The deductibility of a capital loss is different for individuals and corporations.

An individual must first use his capital loss to offset any capital gain; if there is any excess capital loss, he can deduct up to one half of $1000 of his loss from his ordinary income. If after doing this he still has some excess capital loss, he may carry the excess over to future years, repeating this restricted deduction procedure each year until the loss is exhausted.

A corporation cannot deduct any part of a capital loss from ordinary income, but it can use such loss to offset capital gain.

It can carry any excess over for five years to offset future capital gains.

13. Net operating loss carryback/carryover. If your business has an operating loss, or if it has an uninsured casualty or theft loss that exceeds your income, the tax law allows you an offsetting benefit. You may carry the loss back to offset the income of the past three years to gain a tax refund, and you may carry any excess loss over to offset the income of the next five years to reduce future tax.

INVESTMENT CREDIT

The Investment Credit Law became effective on January 1, 1962. It has been modified on numerous occasions since then. Basically this law enables the taxpayer to take a credit against his taxes of from 6 to 10% of the cost of certain property purchased for use in the business for the production of income. The figure of 10% can be changed based on economic conditions. This stimulates the economy by encouraging the modernization of plants. It is a helpful and useful tax law. It is liked by the government since it can be turned on or off either to stimulate or to put a brake on the economy.

In addition to helping the equipment dealer, the credit is a valuable sales tool. Farmers are allowed this tax credit on the purchase of production machinery. Combined with the $5,000 first-year "expensing" deduction of capital assets, it can assure that equipment can be purchased with practically no down payment. Consider this example:

A tractor selling for $30,000 is purchased by a farmer in a 42% tax bracket. He uses straight-line depreciation. The tractor has a life of five years with a $5,000 salvage value.

	Operating expense deduction	Tax reduction
10% Investment credit ($30,000 × 0.10)	0	$3000
$5,000 first year "Expensing" Deduction	$5000	$2100
One-year straight-line depreciation ($25,000 × 0.20)	$5000	$2100
Total first year tax reduction		$7200

A normal down payment would be 25% or $7500. This turns tax credit reduction out to be a powerful sales tool.

DEPRECIATION

Depreciation is an operating expense allowed for the exhaustion, wear, tear, and obsolescence of property used in business. Land is not considered depreciable, nor is good will.

Any asset that has a useful life of less than one year can be depreciated in its entirety the year it is purchased. In the past, assets have been depreciated over their useful life. The value of the asset remaining after its useful life is called the salvage value.

Assets are now grouped into broad categories for purposes of determining depreciation. Fixed write-off periods have been defined. In general, autos, light trucks, and special tools are given a 3-year life. Other equipment, machinery and some farm facilities are allowed a life of 5 years and most buildings, 15 years.

Several methods may be used to compute depreciation. Some methods permit a greater amount of an asset's cost to be written off in the early years of its life. Regardless of the method used, the total amount that can be depreciated is the purchase price of the asset minus its salvage value.

Where capital is limited it is advantageous to use a method that maximizes early write-off. This returns capital to the business so that it can be reinvested. This gives capital without interest cost. Where this is not a problem, a slower rate of write-off may give a tax reduction when income is sure to be greater. This is an individual consideration which should be make with the help of your tax consultant.

Useful life is the estimated length of time you plan to use an asset in your business. This may not be the same as the physical life of an asset. This life is usually estimated from past experience. The Treasury Department publishes guidelines. There are certain limitations, which can be exceeded only with good reason.

If you expect to use an asset for its full serviceable life, the salvage value will be its junk value. Otherwise the salvage value is considered to be its worth when retired from service. The salvage value can usually be disregarded if: (1) you estimate that you will completely exhaust the value of the asset, (2) you use the declining balance method of depreciation, or (3) the asset is personal property with a useful life of at least 3 years and the estimated salvage value is 10% or less of the asset cost.

There are three depreciation methods generally used: Straight Line, Declining Balance and Sum of the Digits. Straight Line is a constant method, while the other two are accelerated methods.

Under the Straight-Line Method equal annual amounts are deducted over the asset's useful life.

$$D = \frac{C - S}{N}$$

where D = Depreciation, dollars per year
C = Cost of assets, dollars
S = Salvage value, dollars
N = Useful life, years

As an example, consider a mobile dynamometer costing $1800 with a useful life of 5 years and a salvage value of $300.

$$D = \frac{1800 - 300}{5} = \frac{1500}{5} = \$300 \text{ per year}$$

Under the Declining Balance Method the depreciation rate is some percentage, always greater than 100, which is multiplied by the amount used in the Straight-Line Method. This higher rate is applied to the asset's remaining cost basis each year. The rate remains constant but the cost declines each year by the amount of the depreciation. A figure of 150% is commonly used for used equipment, whereas 200% may be used for new equipment with a useful life of over 3 years. Salvage value is ignored with the declining balance method except the declining balance of an asset's cost basis cannot be reduced below the salvage value as reduced by 10% of cost.

$$D = \frac{\%(B)}{N}$$

where D = Depreciation, dollars per year
   $\%$ = Percentage of straight-line method
   B = Remaining value to be depreciated
   N = Number of years of useful life

Consider the same dynamometer costing $1800 with a useful life of 5 years and using 150% declining balance.

$$D_1 = \frac{\%(B)}{N}$$

$$= \frac{1.50(1800)}{5}$$

$$= \frac{2700}{5} = \$540$$

$$D_2 = \frac{1.5(1800 - 540)}{5}$$

$$= \frac{(1260)1.5)}{5} = \$378$$

$$D_3 = \frac{1.5(1260 - 378)}{5} = \$264.60$$

The value of the asset is reduced but never reaches zero. For this reason it is permissible to switch over to straight line at any time.

The third method is the Sum of the Digits Method. With this method, depreciation is a fraction whose denominator is a constant (sum of digits) and the numerator is the number of remaining years life times the cost. Using the same illustration, we have:

$$D = \frac{(N - n)(C - S)}{S_d}$$

where D = Depreciation, dollars per year
   N = Useful life, years
   n = Previous times depreciated
   C = Cost of asset, dollars
   S = Salvage value, dollars
   S_d = Sum of digits. This is the sum of all the digits making up the useful life of the machine. There is one digit in one, two digits in two, three digits in three, and so on.

$$D_1 = \frac{(5-0)(1800-300)}{1+2+3+4+5}$$

$$= \frac{5(1500)}{15} = \$500$$

$$D_2 = \frac{(5-1)(1800-300)}{15}$$

$$= \frac{4(1500)}{15} = \$400$$

$$D_3 = \frac{(5-2)(1500)}{15} = \$300$$

This method starts out slower than the Declining Balance Method but is faster in the later years. Like the Straight Line Method, the cost is completely depreciated in the scheduled time.

Other methods may be used if they are reasonable. You may design your own method. Cars and trucks are often depreciated on the basis of mileage. You may use different methods for different items. The reasonableness is usually decided by comparing to the Declining Balance Method.

Table 7.1 compares five different methods. It is based on a machine with a value of \$100 and a useful life of 10 years with no salvage value. The values are rounded off to the nearest dollar. While the annual column shows the depreciation allowed in any individual year, the cumulative column shows the total depreciation allowed to date. This cumulative value multiplied by an individual's income tax rate shows how much and how fast invested capital will be released to

TABLE 7.1. COMPARATIVE DEPRECIATION OF A \$100 ITEM OVER A 10−YEAR PERIOD UNDER DIFFERENT METHODS

Year	Straight Line		150% Declining balance		200% Declining balance		200% D/B Switch to S−L		Sum of digits	
	Annual	Cumulative	Annual	Cumulative	Annual	Cumulative	Annual	Cumulative	Annual	Cumulative
1	\$10	\$ 10	\$15	\$15	\$20	\$20	\$20	\$ 20	\$18	\$ 18
2	10	20	13	28	16	36	16	36	16	34
3	10	30	11	39	13	49	13	49	15	49
4	10	40	9	48	10	59	10	59	13	62
5	10	50	8	56	8	67	8	67	11	73
6	10	60	7	63	7	74	6	73	9	82
7	10	70	6	69	5	79	6	80	7	89
8	10	80	5	74	4	83	6	86	5	94
9	10	90	4	78	3	86	6	93	4	98
10	10	100	3	81	3	89	6	100	2	100

the enterprise for reinvestment. The reinvested capital is in effect interest-free capital available to the business.

The effect of method of depreciation on the resultant tax savings and consequent release of capital to the business can be seen in the following illustration. Values by the year are shown in Table 7.2, with trends shown in the graph Fig. 7.1.

A diesel analyzer has an 8-year life, an acquisition cost of \$3200, and a salvage value of \$200. The taxpayer is in a 40% tax bracket. Declining balance is taken at 150% and converted to straight line on the fifth year. The sum of the digits method releases capital earlier so that it can be reinvested into the business. The graph, Fig. 7.1, further shows that the declining balance method is also preferable to the straight-line method from the standpoint of releasing interest-free capital.

Note: Since texts are static and tax laws constantly changing, check with your accountant or tax advisor before using any of the tax planning suggestions used in this text.

TABLE 7.2. DIESEL ANALYZER DEPRECIATION SCHEDULE[1]

Year	Straight line depreciation ($)	Straight line tax saving ($)	Sum of digits depreciation ($)	Sum of digits tax saving ($)	Declining balance depreciation ($)	Declining balance tax saving ($)
1	375.00	157.50	666.67	280.00	600.00	252.00
2	375.00	157.50	583.33	245.00	487.50	204.75
3	375.00	157.50	500.00	210.00	396.09	166.36
4	375.00	157.50	416.67	175.00	321.83	135.17
5	375.00	157.50	333.33	140.00	298.65	125.43
6	375.00	157.50	250.00	105.00	298.65	125.43
7	375.00	157.50	166.67	70.00	298.65	125.43
8	375.00	157.50	83.33	35.00	298.65	125.43

[1]Based on acquisition cost of $3200, salvage value of $200 and a tax bracket of 40%.

OTHER TAXES

Social Security Tax. This tax is commonly referred to as the FICA (Federal Insurance Contributions Act) Tax. It is levied against both the employer and the employee. The employer must apply for an account number. Each employee must fill a Form W-4, which is a statement of withholding exemptions. On each payment of wages, the employer must deduct the employee's share. The employer submits quarterly on Form 941 both his share and the employee's share. Currently, wages up to $29,700 are taxed at the rate of 6.65% from the employee and 6.65% from the employer. Self-employed persons contribute at the rate of 9.33% on their earnings up to $29,700. Technically speaking, the contribution is not a tax though the effect is much the same.

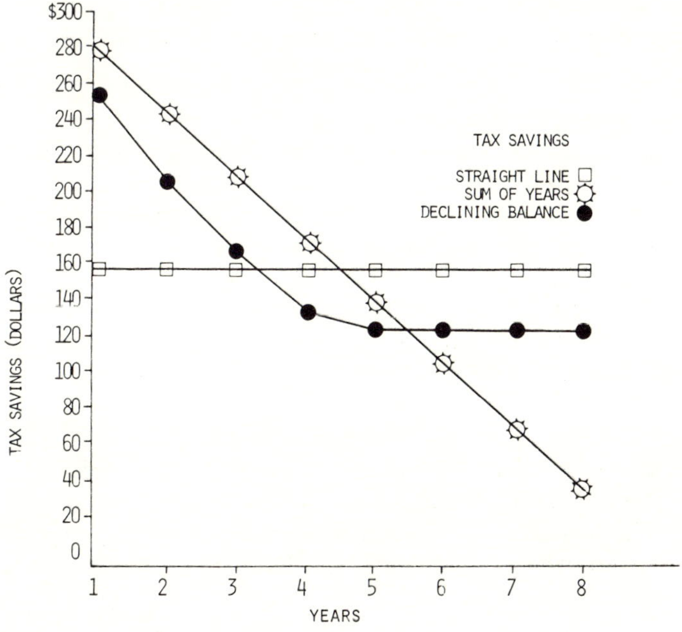

FIG. 7.1. Plot of tax savings when depreciating a $3200 diesel analyzer with $200 salvage value.

Withholding Tax. This tax is entirely on the employee but is collected by the employer and submitted to IRS quarterly. The dealer employer is required to withhold from each payment of wages to an employee an amount representing the federal income tax on the wages paid. Deductions are based on salary and number of dependents. The employee fills out a W-4 form, which certifies the number of his dependents. The tax is remitted directly to the Director of Internal Revenue Service together with the Social Security Tax.

When the sum of the withholding tax and the FICA tax are greater than the current prescribed amount ($100) the payment has to be submitted monthly. Payments are usually made by the 15th of the following month. These payments are made to a Federal Reserve Bank or other designated bank. When quarterly returns are made to IRS, the Reserve Bank receipts are sent in instead of cash.

Each employee must be given a W-2 form in duplicate showing the taxes paid during the year. The triplicate copy of the W-2 form is sent to IRS together with Form W-3, Reconciliation of Quarterly Returns.

Most states also have a withholding tax based on the employees' wages. It is collected in a similar manner and remitted to the State Collector of Revenue. There are wide variations in this procedure from state to state.

Unemployment Tax. There are both a state and federal unemployment tax. This tax is levied on the employer rather than the employee, with the exception of a few states that require a contribution from the employee. The tax is based on the employment experience of the company. The lower the turnover, the lower the tax will be. It behooves all employers to exercise good personnel policies to keep turnover to a minimum.

The tax applies to employers who have four or more employees on at least one day of each of 20 calendar weeks in a calendar year. This requirement may vary slightly from state to state. Corporation officers are counted as employees, but individual proprietors and partners are not.

The Federal Unemployment Tax is paid annually on Form 940. It applies to employers who have four or more employees for 20 weeks.

Federal Income Tax. This tax is felt by employers and employees. It must be taken into consideration in all business matters.

An individual proprietor reports his income from his business on his individual tax return and calculates the tax as an individual.

A partnership is not taxed as such, but an information return must be filed. The return of the individual partners reflects his income from the partnership.

The corporation files its own income tax returns. The tax is calculated according to applicable rates at the time. Individual stockholders pay income tax on their individual returns based on dividends received.

There are certain elections whereby a small corporation may choose to be taxed as a partnership and a partnership taxed as a corporation. The form of organization affects the tax structure of an organization. This as well as salaries paid to owners and stockholders who are employees must be considered in determining the best type of organization.

The state income tax varies from state to state. Generally, the provisions are similar to the Federal Income Tax, but the tax rate is usually considerably less.

State and City Sales Tax.   States as well as cities frequently impose sales tax on merchandise and services sold. This tax is passed on to the customer. Reporting requirements vary from state to state. Retailers should become familiar so that exemptions are not overlooked. The returns are normally filed monthly or quarterly.

Local Taxes.   These take the form of personal property tax, real estate tax, occupational license tax, and others. Rates and methods of collection vary from area to area. It is well to become familiar with local methods to avoid embarrassment.

Tax Planning.   Tax rates are such that the prudent businessman will try to keep them down to a legal minumum. This can be done only during the tax year. Do not wait until the tax year has passed to consider how to accomplish this. Watch year-end items particularly since income can be shifted forward easily then if it would affect taxes. The help of the tax consultant can be profitable.

Arrange tax years to coincide with the time when business will be the slowest. Office help is available without overtime. Inventory is at its lowest. Time is available then to give this problem the maximum attention it deserves.

Record retention is a continuing problem. Statutory limitations must be followed as a minimum. Additionally, it may be prudent to keep some records permanently. Cash books, general ledgers, minute books, and bank statements may be needed long after the statute of limitations has passed. Plan for long-term storage of records in an out-of-the-way place. It's a comforting feeling to know what has gone on in the past.

Tax laws change continually. Rates are adjusted up and down to accomplish certain purposes. Regulations specifying administrative details are constantly being updated. It is wise to let a specialist keep up with this. Trade associations, accountants, attorneys, and management firms all must be relied upon in this area.

LAWS AFFECTING RETAILERS

There are many laws that affect retailers. The brief discussion following will point out those that may have the greatest impact on the farm and power equipment dealer. This discussion is made to create awareness and not necessarily to answer all questions.

Fair Labor Standards Act

This law is often referred to as the Wage and Hour Law. It originated in 1938 and has been amended many times since. The amendment of September 23, 1966, is the one of current interest. It repealed the specific exemption from the minimum wage and overtime provisions of the Act for farm equipment, automobiles, and truck dealers. A special and specific exemption is provided for farm equipment, automobile, and truck dealers regardless of dollar volume of sales from the overtime provision for "any salesman, partsman or mechanic primarily engaged in selling or servicing automobiles, trailers, trucks, farm implements, or aircraft or employed by a non-manufacturing establishment primarily engaged in the business of selling such vehicles to the ultimate purchaser."

Restated, the law says that farm equipment dealers' salesmen, parts men, and mechanics must receive at least the minimum hourly wage, but under the law it is not required that they be paid premium pay for overtime.

Recordkeeping is required by this act. Employers must maintain and keep certain payroll and other records for every employee who is entitled to the benefits of this act.

What Records Are Required?

Sales and Purchase Records.   Every employer is required to keep a record of the total dollar volume of sales or business, and total volume of goods purchased and received during such periods (weekly, monthly or quarterly). These may be kept in such form as the employer maintains in the ordinary course of his business, just so long as they are adequate to establish these volume figures.

Employee Records.   Employers must keep certain records about each worker who is entitled to minimum wage and overtime pay under the federal Wage–Hour Law. The law requires no particular form in the records. All it requires is that the records include certain identifying information about the employee and data about the hours he works and the wages he earns. And, the law says, the information must be accurate.

Here is a breakdown of the information that must be recorded:

A. Identifying Information
1. Name in full, and on the same record, the employee's identifying symbol or number if such is used in place of name on any time, work, or payroll records. This shall be the same name or identifying symbol or number as that used for Social Security record purposes.
2. Home address.
3. Date of birth, if under 19.
4. Sex and occupation in which employed (sex may be indicated by use of the prefixes Mr., Mrs., Ms., or Miss).

B. Hours
1. Time of day and day of week on which the employee's workweek begins. However, a single notation will suffice to cover all employees in the establishment if all have the same workweek, for example, a calendar week. If one employee's workweek differs (for example, Wednesday to Tuesday), a separate notation would have to be made for him.
2. Hours worked each work day and total hours worked each workweek. (For purposes of this requirement, a "work day" is considered as any consecutive 24 hours.)

C. Wages
1. The basis on which wages are paid (such as "$3.90 an hour," "$40 a day," "$160 a week," "$50 a week plus 5% commission on sales over $300 a week").
2. Regular hourly rate of pay for any week in which overtime is worked and overtime payment is due.
3. The amount and nature of any payments that are excluded from the "regualr rate." (These records may be in the form of vouchers or other payment data.)
4. Total daily or weekly straight time earnings.
5. Total overtime compensation for the workweek.
6. Total additions to or deductions from wages paid each pay period. Every employer making additions to or deductions from

wages shall also maintain, in individual employee accounts, a record of the dates, amounts, and nature of the items that make up the total additions and deductions.

7. Total wages paid each pay period.
8. Date of payment and the pay period covered by payment.

What About Timekeeping?

An employer must keep track of the hours worked each day and workweek by every employee entitled to the minimum wage. This is true whether or not the worker also is subject to the law's overtime requirements. It is also true whether the worker is paid by the piece, the hour, the week, the month, or on any other basis.

The employer may use any timekeeping method he wants. He may use a time clock, or have a timekeeper keep track of hours for all workers, or tell the workers to write their own time on the records, or use any other method that suits his needs. If an employee keeps his own time record he should initial or sign each daily entry.

In other words, any timekeeping plan is acceptable, as long as it is complete and accurate. It is the responsibility of the employer to make certain that records are complete and accurate.

If you use a time clock, the number of work hour credits every day and every workweek must be recorded. These figures may be kept on the time card itself or on another record.

What If the Pay Period Is Longer Than a Week?   At some dealerships, the worker is paid every two weeks, twice a month, or for some other period that does not coincide with one workweek. As far as the law is concerned, workers do not have to be paid weekly.

But, because of the law's pay provisions, the employer must keep track of the total hours worked each workweek. That's why it's important that the employer decide the time and the day of the week when the worker's workweek starts.

Remember, you must keep track of work hours on a weekly basis, even when the workweek and the pay period don't coincide. It's easier to keep track of the total hours worked each week if your records clearly separate the number of hours worked each workweek. Work hours must be totaled by the week because overtime pay is figured on the basis of the number of hours worked each week. (Remember that four weeks do not, in most cases, equal a month.)

How Long Should Records Be Retained?

Present Wage–Hour Law provisions require employers to keep all records on which wage and hour computations are based for *two years*, which is the cut-off under the present Statute of Limitations. Records that must be kept for two years are employment and earnings records such as time cards, wage-rate tables, work-time schedules, and records of additions to and deductions from wages. Also keep order, shipping, and billing records for two years.

All of these records must be available for inspection by representatives of the Wage–Hour Division of the Department of Labor, who may request an employer to make extensions, recomputations, or transcriptions. Records may be kept at the place of employment or in a central records office.

The amendments that went into effect February 1, 1967, extended the Statute of Limitations to *three years* in the case of "willful violations." It is, therefore, recommended that all records be retained for *at least three years*.

Posting of Notices

Every employer who is covered by the provisions of the Wage–Hour Law is required to post and keep posted such notices pertaining to the applicability of the law, as shall be prescribed by the Wage and Hour Division. These notices must be posted "in conspicuous places" in every establishment where employees are employed, to permit them to observe readily a copy on the way to or from their place of employment.

Overtime Defined

An employer may establish any workweek he chooses, provided that it is a fixed and regularly recurring period. It need not coincide with the calendar week, but it must begin at the same hour each week. It may be changed only if the change is intended to be permanent. A workweek may be established for an entire business or for different groups of employees in the same enterprise.

The Wage–Hour Law takes a single workweek as its standard and does not permit averaging of hours over two or more weeks. An employee who works 30 hours one week, and 50 hours the next must be paid overtime compensation for the overtime hours worked in the second week.

There is no requirement in the Wage–Hour Law that overtime compensation be paid weekly. Overtime compensation earned in a particular workweek must be paid on the regular pay day for the period in which such workweek ends.

Overtime begins when 40 hours have been worked in a week. It has nothing to do with hours worked in a day or with work done on Sundays or holidays. For example, a bookkeeper employed by a dealer might work 10 hours a day for four days without becoming eligible for overtime. However, any work done by him on the fifth day of that week would be paid for at his overtime rate.

Remember again, however, that overtime pay requirements of the Wage–Hour Law do not apply to salesmen, partsmen, and mechanics engaged in selling or servicing farm equipment, automobiles, or trucks.

Provisions for Student Employees and Handicapped Workers

A special amendment to the Wage–Hour Law provides that student employees and handicapped workers may be paid at wage levels below the normal established minimum hourly rates as follows:

Student Employees—Not less than 85% of the minimum hourly rate in effect at the time for employees in that establishment.

Handicapped Workers—Not less than 50% of the minimum hourly rate in effect at the time for employees in that establishment.

In order to pay these lower hourly wage rates to student employees and/or handicapped workers, the dealer must apply for and receive certificates permitting him to do so from a Department of Labor office.

Wage and Hour inspectors are entitled to examine your records and discuss your operation with employees. You can accompany an inspector during the investigation. You are required to furnish him any information he may want, but only after he makes a formal request in writing. A cooperative attitude is always helpful. It is usually not wise to volunteer any information not asked for.

Federal Trade Laws

Several acts have been passed to regulate trade. They are administered by the Federal Trade Commission, an arm of the Department of Justice.

The Clayton Anti-Trust Act of 1914 was passed to prevent price differentials to various dealers. This was an attempt to curb monopolies and prevent unfair competition. The Robinson–Patman Act amended the Clayton Act to provide that price differentials, including quantity discounts, must be based on differences in grade and quality or actual differences in cost of manufacturing or delivery. An exception allowed differentials in individual cases where it can be shown that the only intention of the price differential by a seller is to meet competition.

Some companies have attempted to dodge the provisions of this act by offering the same quantity discount to all purchasers. The quantity was set so high that only a few large dealers could handle the order. This resulted in the Federal Trade Commission stepping in and setting quantity limits. In the case of automobile tires, the quantity is limited to one carload. Other discounts have been limited as well. Generally speaking, any discount offered to one must be offered to all retailers on the same basis.

The Federal Trade Commission is here to protect the rights of all. Any retailer who feels he is being discriminated against has the right to complain to the FTC. Action is usually quicker if more than one retailer makes the complaint. The FTC has the power to issue a "cease and desist" order. Make certain of your facts before complaining.

Exclusive Contracts.   The Clayton Act does not prohibit a manufacturer from entering into an exclusive contract to sell products through one dealer in a given area or to confine the territorial limits of a dealer. A manufacturer may not force a dealer to sell only his line to the exclusion of other products. This generally is considered to be in restraint of trade.

A manufacturer can require a dealer to maintain a place of business adequate to serve the public. The dealer cannot be forced to build a new building to certain specification with threats of franchise cancellation.

Price fixing between producers, wholesalers, or retailers is generally an illegal restraint of trade under the Sherman Anti-Trust Act or an unfair practice under the Federal Trade Commission Act as well as a violation of most state anti-trust laws.

Labor Laws

The Taft–Hartley Labor Management Relations Act of 1947 is the basic act that prescribes the rights of employers and employees in labor relations. The purpose of this act as declared by Congress is to provide orderly and peaceful procedures for preventing interference by either with the rights of the other, to protect the rights of the individual employees in their relation with labor organizations whose activities affect commerce, to define and prescribe practices on the part of labor and management that affect commerce and are inimical to the general welfare, and to protect the rights of the public in connection with labor disputes affecting commerce.

The act was designed with big business in mind. The National Labor Relations Board has jurisdiction. The McClelland Committee, as a result of its study, was responsible for the Labor–Management Reporting and Disclosure Act of 1959. Its report concluded that most unions and businessmen operated with honesty and integrity. In the course of the investigation the Committee did find widespread abuses in certain areas and reached the following conclusions:

1. There has been significant lack of democratic procedure in the unions studied.
2. The international unions surveyed have flagrantly abused their power to place local unions under trusteeship or supervisorship.
3. Certain managements have extensively engaged in collusion with unions.
4. There has been widespread misuse of union funds.
5. Violence in labor–management disputes still exists to an extent where it may be justifiably labeled a crime against the community.
6. Certain managements and their agents have engaged in a number of illegal and improper activities in violation of the National Labor Relations Act.
7. The weapon of organizational picketing has been abused by some of the unions studied.
8. Gangsters and hoodlums have successfully infiltrated some labor unions, sometimes at high levels.
9. An extensive "no man's land" in labor–management relations has been uncovered by committee testimony.
10. Law enforcement officers have been lax in investigating and prosecuting acts of violence resulting from labor–management disputes.
11. Members of the legal profession have played a dubious role in their relationships with officials of some unions.

On the basis of these conclusions, legislation was recommended by the committee to regulate and control pension, health, and welfare funds; regulate and control union funds; insure union democracy; curb middlemen in labor–management disputes; and clarify "no man's land" in labor–management relations. The labor–management reform legislative proposals were introduced during the second session in the 85th Congress but were defeated in the House after passage in the Senate. In 1959, the principal recommendations of the Committee were enacted into law during the closing days of the first session of the 86th Congress.

Control Extensive.   This new "Labor Law of 1959" was heralded as another milestone in the effort to achieve an equitable balance in the field of labor–management relations. New controls affecting labor unions and their relationships with union members were sweeping and inclusive. A "bill of rights" was established for union members covering participation in union affairs. The new labor law made extensive and important amendment to the Taft–Hartley Act as to extortion, picketing, economic strikes, secondary boycotts, recognition of picketing, and hot cargo contracts. Certain changes affecting the jurisdiction of the National Labor Relations Board were also included.

Unions and This Industry. By far the greater segment of the retail farm and power equipment industry has never become greatly concerned over the possibility of the "union shop" within their own stores. Labor unions are spreading their area of influence as each year passes. Ahead of them lies a vast untouched field of retailing. Ever-increasing pressure to extend coverage under the Wage–Hour Act only serves to hasten the day when labor unions will attempt to encompass the field of retailing.

The only defense for the retail farm and power equipment dealer is to perfect his relationship between himself and each and every one of his employees. This is of great importance.

Truth in Lending Law

In 1968 Congress passed the "Consumer Protection Act," which is referred to as the "Truth in Lending Law." It became effective on July 1, 1969. The law does not limit the amount of interest that can be charged, but it does require that you clearly and conspicuously inform your customers of your credit provisions.

The law is enforced by the Federal Trade Commission, but the regulation (Regulation 2) was issued by the Federal Reserve Board. A retailer is concerned primarily with conditional sales contracts, promissory notes covering purchases of equipment, and open accounts where a service charge is made if an account is not paid in time.

The principal provisions of the law require that you do the following:

1. Prior to extending credit to a new customer you must mail or deliver a written notice of your credit provisions.
2. Your monthly statement form must show previous balance, finance charge payments, credits, and new balance as well as other headings you may personally need.
3. If credit provisions are shown on the shop ticket or sales tickets, they must state clearly and accurately the rates, free time to pay without charges, minimum finance charges, and so on.
4. In open account credit you must use the term "finance" rather than interest, service charge, or carrying charge.
5. Your printing must be of a minimum size. All amounts or percentages must be 10-point type or larger.
6. On open account credit you must state the finance charge as an annual percentage rate.

Employee Retirement Income Act of 1974

This law regulates about 300,000 pension and welfare plans. It does not require a minimum level of benefits, but it regulates extensively the substance of all existing and future plans. It not only deals with safeguarding money contributed to plans, but also with how the money is accumulated, how it is distributed, who controls it, and what happens if there isn't enough.

All pension, profit-sharing, thrift, and savings plans must guarantee future retirement benefits to participants who complete a specified number of years of service, even if the employee leaves employment before reaching retirement age. This vesting requirement consists of any of three optional plans.

1. Graded vesting—vesting of 25% of accrued benefits after 5 years' service, increasing by 5% a year to 50% after 10 years' service, then by 10% a year to 100% after 15 years.
2. Ten-year formula—100% vesting after 10 years' service.
3. Rule of 45—50% vesting when age and service equal 45, increasing by 10% a year to 100% five years later.

Paying vested benefits may be made according to the plan and begin at age 62 but no later than age 65. Benefits are payable in full at that time. A vested pension may not be suspended after normal retirement age because an employee is guilty of misconduct or is subsequently employed by a competitor.

Every pension plan must give participants the option of providing continuing benefits to a spouse in the event the retired employee dies before his or her spouse. The amount provided must be at least 50% of the employee's pension. The cost of the option may be borne by reducing the employee's benefit.

Every defined benefit must have certain funding standards. All plans must have their cost calculated by an actuary at least once every three years. The annual contribution to the plan must meet the normal cost. The law sets up a Pension Benefit Guaranty Corporation made up of the Secretaries of Labor, Treasury and Commerce, to insure pension plans in case the plan terminates without sufficient funds to meet its vested obligations. This insured benefit is limited to $750 a month.

The law provides for individual retirement accounts (IRA's) to be established by an individual or their employer or unions. An employee who is not covered by a plan may contribute up to 15% of his earnings, not to exceed $1500 a year, to an "individual retirement account." This contribution is tax exempt. This arrangement may permit an employer to cover particular employees not otherwise covered. The employer may make part or all of the contribution. This is considered income to the individual but deductible by the employee for income tax purposes.

Total annual employer contributions to a profit-sharing plan are limited to 15% of payroll. If less is contributed during one year, the contribution on subsequent years may be increased up to 25% of payroll to make up the difference. Annual contribution on behalf of an individual may not exceed $25,000.

The law is massive. The Treasury Department through the Internal Revenue Service has jurisdiction over qualification of employee benefit plans including vesting, funding, and many fiduciary standards. The Labor Department has jurisdiction over fiduciary standards, disclosure, and defining service years and breaks. It has the power to grant temporary vesting variances and extension of amortization periods and to investigate plan operations. The law went partially into effect in 1975 and was fully in effect early in 1976.

During the 1981 legislative session, Congress enacted the "Economic Recovery Tax Act of 1981," which further encouraged the use of private pension plans. People who are covered by an employer-sponsored retirement plan may also open an individual retirement account offered by banks, savings and loans, mutual funds, and other financial institutions. The maximum contribution is $2000 or $2250 if split between accounts for an individual and a nonworking spouse. The amount cannot exceed the yearly compensation. If the employer agrees, an individual may also contribute to a company pension plan with those payments qualifying for IRA benefits.

Persons not covered by an employer's plan may also contribute up to $2000 to an individual account or $2250 if the account is opened with a nonworking spouse. Contributions may be as much as 100% of compensation.

Self-employed persons may set aside earnings in a Keogh plan similar to an IRA. They may contribute and exclude from current taxation up to 15% of their earnings to a top contribution of $15,000. This was raised from the previous top of $7500.

Occupational Safety and Health Act

Public Law 91–596, the Occupational Safety and Health Act of 1970, was passed with the purpose of providing safe and healthful working conditions for employees. It is administered by the Department of Labor. States may make a state plan, get it approved, and enforce it. The federal government will partially finance the cost of administering the plan.

The act opens with these words, "To assure safe and healthful working conditions for working men and women; by authorizing enforcement of the standards developed under the Act; by assisting and encouraging the States in their efforts to assure safe and healthful working conditions; by providing for research, information, education, and training in the field of occupational safety and health; and for other purposes." That's the way the Williams–Steiger Occupational Safety and Health Act of 1970, administered and enforced by the U.S. Department of Labor, announces itself.

It continues by stating, "The Congress finds that personal injuries and illnesses arising out of work situations impose a substantial burden upon, and are a hindrance to, interstate commerce in terms of lost production, wage loss, medical expenses, and disability compensation payments."

Congress thus takes a major step in forcing the adoption of uniform working conditions, everywhere in the country, for every person who is employed. And who must comply with the requirements of the Act? Basically, everyone. Farmers who employ 10 or fewer people are exempt from inspections and record keeping. Small businesses who employ 10 or fewer are subject to inspections but are exempt from the recordkeeping requirements. Employers with 20 or fewer employees guilty of violations are fined at a 40% reduced rate.

Prior to the effective date of the act, every dealer should have received a booklet, "Recordkeeping Requirements." You must make every effort to comply with the requirements of the act.

1. The centerpiece of the booklet is a poster, which must be displayed in a prominent place in every dealership. It must be in the area in which employees normally work.
2. A log of occupational injuries and illnesses, OSHA Form No.

100, is the next form in the booklet. Each recordable occupational injury or occupational illness must be entered on the log within six days of receiving information that a recordable case has occurred.

3. A summary of occupational injuries and illnesses, OSHA Form No., 102, must be prepared at the end of each year within one month following the end of that year. The summary should be posted in the place accessible to employees (with the poster). It must go up for a period of 30 consecutive days within one month after the close of the calendar year.

4. A supplementary record of occupational injuries and illnesses must be kept for each recordable injury or illness. Most Workmen's Compensation Insurance reports will be acceptable as a substitute for OSHA Form No. 101.

Compliance officers (inspectors) representing the Department of Labor will inspect your dealership for violations of OSHA safety standards. They will concentrate first where accidental deaths have occurred, and such a death must be reported to the Secretary of Labor within 48 hours.

Complaints from your employees, not only current ones, but past ones, too, can bring the inspectors to your dealership. Considerable time is being devoted to what OSHA calls "target industries," but about 25% of the inspector's time is going to spot checks.

If an inspector (compliance officer) should visit you, first be courteous. Be considerate and cooperative and show good faith in trying to promote safety. He most certainly will be able to find something to criticize. If you receive a citation for a violation you will be given a specified period to comply (called an abatement period). You must correct the situation within the abatement period or be subject to even more severe fines. You will not gain a thing by being disagreeable. In fact, it will probably cost you added grief.

Now, what will the inspector look for? Here's a possible list:

Fire Extinguishers—A good test would be to ask a stranger to walk into your shop and quickly find them. If they are difficult to locate, you have a problem. They should be hung on the wall, not more than 5 feet above the floor; should be easily visible; and should be tagged every year and be of an approved type for the kind of fire they are expected to combat. The color code for fire protection equipment is red. A red background against which to hang your extinguishers can call immediate attention to their location.

Sanitary Conditions—Some inspectors put this condition high on their list. They want to see the washroom equipped with hot and cold water, a covered trash container, and no food or lunches stored therein. Naturally, all areas of the building should be clean, with aisles unobstructed and wide enough to accommodate the traffic. Toilet facilities and drinking water must be available within 200 feet of where employees normally work. Each sex must have a separate toilet.

Lighting—A majority of citations have been issued for inadequate lighting in basement stairways. Brighten them up.

Spray Painting—Here you face almost certain problems. Spray paint rooms must be properly ventilated around all spray areas and be equipped with explosion-proof lights, automatic fire extinguishers, and no-smoking signs. All motors should be of the nonsparking kind. Some who have studied the standards applying to paint operations have said, "Unless you have a spray facility that meets every OSHA standard, do not paint inside. Paint outdoors or contract the work to someone else."

Tools—Portable power tools must be grounded unless double insulated. Building wiring, naturally must be able to accept three-prong plugs. Hand tools may not have "mushroomed" heads. Turn your air compressors so the belts are to the wall or else install strong guards. The air hose must have a safety nozzle and must not allow cleaning air to blast forth in excess of 30 psi. Grinders? All guards and tool rests MUST be in place and be USED. Power tools must not have automatic "on" locking trigger.

Personal protective equipment—Men using grinders must wear safety glasses, and welders must have and wear goggles and gloves. Printers must wear face masks. If you have employees working where sound levels exceed 90 decibels for 8 hours continuous exposure, they must have ear protective devices (ear muffs). Employees on construction sites, for delivery or repair work, MUST have "hard hats."

Some additional suggestions are first aid kits should be kept well stocked. Check with your physician for his stocking suggestions. A box of Band-Aids won't suffice.

Ladders must be in top condition. Get rid of those shaky relics right now. Just standing around they could cost you a citation even if they are never used. Straight ladders must have safety feet.

Gasoline cannot be drained into open containers and it cannot be used for cleaning purposes in the shop. It can only be stored in safety containers with spring-loaded, quick-release pour spout caps. A screen-type flame arrester must be in place in the pour spout, too.

Stairways must be capable of carrying a load of five times the normal live load and not less than 1000 pounds moving load. All stairs must have railings, and bright lighting is a necessity.

Decks and loading docks require protective railings, and a 4 inch toeboard should be in place. Guards or chains should be around high-level (4 feet or higher) storage areas to prevent objects from falling and all ramps to these docks must have guard railings.

Hallways, storerooms, service shops and all work areas must be clean, orderly, and sanitary; all exits should be visibly marked. The safety standard requires "EXIT" signs to be lighted with a "dependable" light source.

Drains and gutters, covered with gratings, must be installed to drain all potentially wet areas of your shop.

Most of the recommendations are actually just good safety suggestions, developed over many years of experience. These suggestions are just a small portion of the total safety requirements. A few could probably be classified as "nit-picking," but safety people and industry representatives are working to get desired changes in codes as fast as possible. In the meantime, you are to comply with existing safety standards as specified by OSHA.

Remember, your attitude should be to make every effort to provide a safe place of employment for your crew. If you act in good faith and cooperate with OSHA representatives, you will have a minimum of problems.

How Complaints Are Handled

Employees may make oral (during time of inspection) or written complaints to OSHA if they feel they are working under hazardous conditions that need correcting to comply with some existing safety standard.

Their names do not have to be revealed to the employer. In fact, in this connection the Act is quite specific in pointing out that you may not discharge or in any manner discriminate against any employee because he files such a complaint against you. There are teeth in the Act. Any employee who believes that he has been discharged or otherwise discriminated against in connection with his complaint may, within 30 days, file a complaint with the Secretary of Labor. The Secretary's representative is then authorized to investigate the matter and to bring action in the U.S. District Court for appropriate relief, including rehiring of the employee to his former job with back pay.

A careful reading of the Act and material connected with it quickly shows that the "rights" are heavily weighed toward employees. As an employer, your rights in the matter are few. Full responsibility for compliance rests with you.

Since OSHA places specific responsibilities on each employer, one of which is to enforce the safety standards in your own place of business, here is a provision of which you need to be aware. NO obligations are placed upon the employee by the Act, only upon you.

The employee will not receive a federal citation or a penalty if he fails to follow safety orders. The burden of responsibility is on the employer's shoulders for complete enforcement of all federal safety standards.

You will be wise if you make it a condition of employment at your dealership that all employees must comply with all OSHA regulations and that failure to do so is sufficient cause for dismissal.

Remember, it need not be a current employee who files a complaint against you. A disgruntled, former employee can cause you extreme grief when he "suddenly remembers" some phase of your operation that he considers is still "unsafe" for employees.

How to Appraise Your Problems

First, one thing is quite obvious. No matter what you do to try to comply with OSHA's regulations, it is always possible that you will draw a citation (fine) for something when the inspector calls. Much is left up to individual judgment when it comes to determining whether a situation is hazardous enough to demand that a citation be issued or not.

It's harder for you to personally see problems around the store, for you have lived with its conditions so long that they will seem natural. It is a good idea to call on some outside help from time to time.

As you develop your voluntary compliance program, invite your insurance man to come in and give your place a thorough inspection. Note his suggestions and then make corrections.

Call your staff together and enlist their aid in complying with OSHA, for it can literally mean their jobs are at stake. If they don't comply with legal safety regulations, they are eligible for dismissal (and you for a fine). Implementing OSHA regulations will mean increased costs for you, and employees should be made aware of that, too.

Then, after you have put into effect all the steps suggested by your insurance people and by your own employees, after you have studied OSHA standards you can institute a complete clean-up, fix-up program around your dealership. By doing this, you will have gone a long way toward reducing the number of citations you will draw when the inspector calls. Work to eliminate the obvious safety hazards first and fast.

Important Standards to Study. For your dealership to comply with OSHA standards, you need to know what those standards call for. The following are taken from the Federal Register, Volume 36, Number 105, dated May 29, 1971. You should have a copy. You may obtain a copy from the Government Printing Office or the nearest OSHA office.

Below are safety items taken from the OSHA standards, which will apply to most dealerships. Refer to the indicated section for complete details.

1. Display room, parts department, warehouse, and service shop must be kept clean and orderly and in a sanitary condition. (Sec. 1910.22)
2. All floors including the service shop, must be kept clean, and so far as possible, dry. Proper drains must be located in both shop and wash room. (Sec. 1910.22−2)
3. Eliminate from every floor and passageway all splinters, holes, or loose boards and protruding nails. (Sec. 1910.22−3)
4. If you use mechanical handling equipment, allow for safe clearances at loading docks, through doors, and between any aisles. You should also mark permanent aisles with paint. (Sec. 1910.22−3B)
5. Covers and/or guard rails shall be provided around open pits, tanks, and ditches. (Sec. 1910.22−3C)
6. Don't overload your floors. This applies particularly to floors that are not poured directly on the ground. Parts room balconies and other upstairs rooms must be checked by a building inspector and then be posted with signs that indicate maximum floor loads allowed. If you exceed those maximums, you are liable, for the building inspector's maximum is your floor's legal load limit. (Sec. 1910.22-d)
7. Floors above ground level are given special attention. Remember every stairway to upper levels must have a standard railing. Balconies must have railings and toe rails. If you have a door opening out into the service shop from a second-floor room, it must be guarded by a rail, roller, picket fence, or half door. The guard may be removable, but it should preferably be hinged. A removable toe board is also called for. (Sec. 1910−23 A1)
8. A balcony over your parts department or offices must be protected by a standard railing and 4-inch-high toe board on all open sides. (Sec. 1910−23B)
9. Every flight of stairs in your building having four or more risers shall be equipped with standard stair railings or hand rails. On steps less than 44 inches wide, one hand rail is required on the right side, descending. Stairs wider than 44 inches need one rail on each side. If stairs are open, there must be a railing on both sides. (Sec. 1910−23D)
10. If you have fixed ladders in your building, the minimum design live load they must support is 200 pounds. Mobile

work stands your mechanics use in the shop must safely carry a minimum load of 250 pounds, one person occupying them at a time. When more than one man is on them, they must use platforms with greater load capacity. They shall be capable of supporting at least four times the design work load. (Sec. 1910—27)

11. All exits must be kept clear at all times. And repeating, they must be marked by a suitable illuminated, reliable light source at all times. The signs must have the word EXIT in plainly legible letters not less than six inches high. (Sec. 1910—37)

12. Your spray paint operation is a particular problem. Suffice it to say that only the finest, well-designed, commercially built, professional installed commercial spray paint booths can possibly comply with this standard. You will probably be better off to have someone with such paint facilities do the job for you. By law, spray painting as it is done in many dealerships is illegal. (Sec. 1910—94 and Sec. 1910—107)

13. You must provide eye and face protective equipment for your employees (or if they provide their own, you are still responsible for its condition) if there is reasonable probability that injury can be prevented by such equipment. This means welding goggles, safety glasses at the grinders, and even possibly protective glasses for servicemen working under machines from which dirt or dust may fall in their eyes, and the men MUST wear them. It's the law. (Sec. 1910—132)

14. And speaking of the law, it's now a violation of federal law to expectorate (spit) on the walls, floors, or stairs. (Sec. 1910—141—2)

15. No water cooler you provide may allow the ice to come in contact with drinking water, and drinking water must be available within 200 feet of any location at which your employees regularly work. (Sec. 1910—141B)

16. You must provide separate toilet facilities for each sex, and they are to be readily accessible. Construction specifications are quite strict and very specific about toilet facilities. Better read Federal Register, Vol. 36, No. 105, Subpart J, General Environmental Controls, page 10593, for full details.

17. Fire extinguishers, fully charged and correctly inspected, must be in easy-to-see spots. They should be of types designed to extinguish the type of fire most apt to ignite in the area they protect. For instance, an extinguisher for fires in flammable liquids and greases should be in the shop, while an extinguisher for combustible material such as wood, paper, and rubber is needed in the parts department. Don't set the extinguisher on the floor where it may be hidden. Hang it properly. (Sec. 1910—157)

18. Shop cranes and overhead hoists will come in for close study by the inspector. Thirteen pages of fine print in the Federal Register outline how such cranes will be used, the loads they must support, and how they will be maintained. If you use such cranes, better check page 10617 of Federal Register, Vol. 36, No. 105.

19. Your machine shop equipment must have all guards in place to protect the operator and other employees in the machine

area from hazards such as those created by its operation. (Sec. 1910–212)

20. You are responsible for the condition of all tools and equipment used by your employees. This includes a section prohibiting the use of compressed air in excess of 30 psi for any cleaning purposes—and then only with effective guarding and personal protective equipment such as goggles or safety glasses. (Sec. 1910–242)

21. Do not use old, junk mowers to cut the grass around the dealership. Only the best of new mowers, complying with OSHA safety standards, may be used by your employees. (Sec. 1910–243–E)

22. Any jacks used by your servicemen must have the rated maximum load legibly marked in a prominent location on the jack. Jacks will be inspected once every 6 months, and jacks that are broken must carry tags that warn against their use. (Sec. 1910–244)

23. Your welding-gas tanks must be stored at least 20 feet from any highly combustible material and away from the stairs or elevators. Check the 12 pages on welding starting on page 10656 of the Federal Register, Vol. 36, No. 105. (Sec. 1910–251)

If You Are an Industrial Dealer.   If you are selling, or renting to contractors, there are some additional specific standards of which you need to be aware; however, most of these also apply to ag dealers. These particular ones are condensed from the Federal Register, Volume 36, No. 75, dated April 17, 1971.

For instance, you should protect employees against the effects of noise exposure. When employees are subject to sound levels in excess of 90 decibels for more than 8 hours, they must wear personal protective equipment—ear muffs. (Sec. 1518–50)

Drivers delivering equipment to construction sites must wear hard hats. (Sec 1518–100)

Make certain all safety belts are in place on machines as required. (Sec. 1518–104)

If you do tire work, a safety rack, case, or equivalent protection must be provided and used when inflating, mounting, or dismounting tires installed on split rims or rims equipped with locking rings or similar devices. (Sec. 1518–600)

Applicable, whether you are primarily selling industrial or farm, is a section regarding your motor trucks. It is unlawful to allow an employee to drive a vehicle with cracked and broken glass. All vehicles shall have defrosting or defogging equipment. Seat belts must be in place for all employees transported and employees must use them. (Sec. 1518–601)

If you have a tilt bed truck, the operating levers must be equipped with a latch to prevent accidental starting or tripping of the tilt mechanism. (Sec. 15180601)

Another section says that all rubber tired motor vehicles manufactured on or after May 1, 1972, shall be equipped with fenders. All units manufactured before that time must be fitted with fenders before May 1, 1973. You can use mud flaps only when the motor vehicle was not designed for fenders. (Sec. 1518–601–13)

All bi-directional machines such as rollers, front-end loaders, bulldozers, and similar equipment must be fitted with a horn, loud enough to be heard over surrounding noise, to be operated as needed when the machine is moving in either direction. If the machine has an obstructed view to the rear when it is in reverse, a signal alarm must be sounding. (Sec. 1518−602−9)

The law is rather new and untried. It is here to stay. Look for standards and procedures to be changed based on experience. Your trade association will keep you up to date with new developments.

Hazardous Waste

The Environmental Protection Agency (EPA) is responsible for enforcing certain standards. Many states have in turn accepted this responsibility and passed appropriate legislation. There are two areas where dealers may become involved, pesticides and oil wastes.

The disposal of pesticides (herbicides, fungicides, nematacides, insecticides, etc.) and their containers is now regulated in most states. The disposal of waste wash water when repairing used sprayers will pose problems. Guidelines published in "Apply Pesticides Correctly" by the U.S. Environmental Protection Agency and the U.S. Department of Agriculture should be helpful.

Most states have adopted regulations regarding the disposal of used motor oil and solvents. Consult local authorities for proper procedures.

QUESTIONS

1. What do we mean when we say a corporation has perpetual life?
2. Distinguish between accrual basis and cash basis.
3. How can investment credit be used as a sales tool?
4. Which depreciation method gives you the fastest tax write-off?
5. What are the main provisions of the Fair Labor Stands Act?
6. What trade laws have a bearing on the farm equipment retailing business?
7. What is the purpose of the "Occupational Safety and Health Act?"
8. How does OSHA affect the farm equipment retailing business?
9. Under OSHA, who is responsible for seeing that employees wear suitable protective equipment?

Business Records and Procedures

Records are the basis for management decisions. They are of interest as historical documents, but their prime purpose is to assist in charting the future and controlling the present. They point to hard decisions that must be made. To the extent that these decisions are made and carried out, recordkeeping will be a boon to management. If records are kept to chart the past and gather dust in the safe, then they become just another cost of doing business.

BASIC RECORDS

Records provide the balance sheets and operating statements that are used to plan for the future, to correct wasteful practices, to denote trends while they are in the process of happening, to manage cash flow, to keep inventories in balance and turning over, and to keep all phases of the business in the proper proportion to each other. The balance sheet is a statement of:

1. What the business owns (assets)
2. What the business owes (liabilities)
3. What it is worth on a specific date (net worth)

It classifies the manner in which capital is invested, what obligations are currently due and must be paid from current earnings, what obligations can be paid from future earnings, and the gain or loss of net worth.

The operating statement often called a Profit and Loss Statement shows:

1. What the business sold
2. Gain or loss over the cost of goods sold
3. Expenses for operating the business
4. Remaining profit or loss

The balance sheet tells management the condition of the business at a given date and the Profit and Loss Statement tells how successfully the business has operated in the last period.

Additionally, records are used to maintain customer lists and customer accounts, to maintain personnel records, to process payrolls, to pay accounts, to generate financial ratios, and to record profit control plans. This list could go on and on.

Records are not a guarantee of financial success. The manager must learn how to evaluate the records. Even more important, he must be willing to do what the records indicate must be done. Many times this may turn out to be disagreeable. Consider the manager who finds that his service shop is selling only 50% of its labor. He is faced with the problem of telling one mechanic that his services are no longer needed. Everyone pictures himself as the manager who passes out raises and bonuses to faithful employees. He must also be the manager who keeps the salesman's margins comfortably greater than the cost of doing business—even though it means reduced sales.

THE BOOKS

The books refer to an accumulation of accounting forms that are used to record data for the business. They may be maintained by hand or by electronic data processing equipment or a combination of both. They may be sent out to an accountant or to an electronic data processing center for analysis, or they may be completed at the business site by the bookkeeper. The method to choose is an economic or convenience factor. Either method can be quite satisfactory. The computer method will save one-fourth to one-half of the bookkeeping time.

There are a number of companies who specialize in the accounting business. It is usually advantageous to deal with one that specializes in accounting for the farm and power equipment business. Several of the major line manufacturers offer this service to their dealers. The National Farm and Power Equipment Dealers Association offers accounting services to their members through the Dealers Computer Accounting Manual Input System (DCA-M).

Dealers Computer Accounting Manual Input System

In this simplified system designed for the farm equipment dealer, the dealer posts daily business transactions on easy-to-use duplicate journal forms. The original copy of the journal is mailed to the Association and in turn a complete set of books is returned to the dealer each month. The following forms are produced, in balance, ready to use:

1. Proof and Balance Report
2. Customer Statements
3. Aged Analysis of Accounts Receivables
4. General Ledger (detail entry)
5. Balance Sheet
6. Operating Statement
7. Expense Statement
8. Department Profit and Loss Statement
9. Monthly Comparative Analysis

The system is based on input journals kept by the dealer. The forms are double-entry balancing and are representative of forms bookkeepers are already familiar with. They include the following:

1. Daily Charge Sales and Receipts
2. Daily Cash Sales
3. Whole Goods Sales
4. Disbursements
5. Purchases—Accounts Payables
6. Cash Paid Out and Deposited
7. Payroll
8. General Entries

From these the following information is compiled and returned to the dealer monthly.

Proof and balance report. This report verifies that the handwritten entries submitted agree completely with the report. It lists all accounting entries in page number groupings as submitted on journal forms by the dealer. The Proof and Balance Report lists all amounts in debit and credit columns, totalled by journal page, and grand totalled for the month to date. The count of entries from journal pages is similarly accumulated and printed at the front of the report.

Customer statements. Monthly statements of accounts prepared in the DCA−M System are pre-addressed and printed in duplicate. The original copy is separated and ready to insert into standard, small-size window envelopes. The remaining copy is for your records. The statements are returned in bulk to the dealer. The dealer then checks the statement, attaches the charge slips, and mails them to the customer. There are various forms of the statement available according to the way finance charges are to be handled. Additionally, the statement contains an aging report on accounts receivable activity for the customer. Accounts receivable are reconciled for accuracy and the report sent to the dealer each month with the statement and aging reports.

Aged analysis of accounts receivables. This report lists all customers having a balance account as of the date of the report. Aging starts with a current column and proceeds to past due accounts of one, two, and three months and over. A column also reports finance charges that have been added to the account, last date customer paid on his account, and the last amount paid.

Preliminary trial blance and general ledger. These reports are identical in design and in programming. The preliminary report is ordered whenever additional activity is planned or expected for the month. Normally only the general ledger is used. The ledger shows beginning balances of general ledger accounts, detail activity for the current month, the date and source of current entries, and the ending balances. Each item in the current month entries column originates from a specific journal page which is shown. Footings of this report show profit and loss year-to-date, total debit and total credit, and a summary of page total debit and credit.

Financial reports. The conventional balance sheet, operating statement, expense statement, and departmental profit and loss statement are computed each month. The operating statement shows the results and the sources of profit or loss for the business. It lists sales and cost of sales accounts side by side, with gross margin calculated for each category and totaled by departments. The report covers both current month and year-to-date. The expense report lists all of the detailed operating expenses and shows percentage relationship of each account to total sales, current month, and year-to-date. This report gives you all the information you need for control, every month. The department profit and loss statement is a very important managerial tool. It shows what each department is doing. Expenses are allocated to the departments by percentage, which is assigned by the dealer. Sales, cost of sales, margins and net profits are computed for the current month and year-to-date.

Monthly comparative analysis. Dealers who participate in DCA−M are automatically included in the annual and mid-year Cost of Doing Business Survey. Comparison reports cover the balance sheets, operating statement, and expense statement. The dealer's own figures are compared to dealers in the same region, dealers handling the same brand, and dealers in the same volume category. Significant ratios are computed such as inventory turnover, percentage of ownership, margin and profit percentages, and expense percentages related to sales.

FIG. 8.1.   The computer terminal is a valued friend to the bookkeeper.

THE BOOKKEEPER

Although the bookkeeper's job is greatly reduced when electronic date processing is used (Fig. 8.1), it isn't eliminated by any means. The input journals previously described must be prepared.

Where all books are kept locally, the job is quite voluminous. The following timetable taken from the National Farm Equipment Dealers Association and based on their FER500A System may serve as a reminder of the many details that must be kept up with. Details may vary, but the basic procedure would be the same.

Daily Bookkeeping Time Table
  Sales and Receipts
  1. Enter cost figures on sales tickets for whole goods and service labor (by reference of invoices and repair orders).
  2. Enter all sales figures for sales, receipts, and cash paid out on the Daily Sales and Receipts Summary.
  3. Post charge sales tickets to Customers Ledger.
  4. Post sales tickets for payments received on account to Customers Ledger.
  5. Balance the Summary, and summarize the several miscellaneous columns.
  6. Balance the cash.
  7. Post the Daily Sales and Receipts Summary totals to the Sales and Receipts Record.
  8. Post "Cost of Sales" column to Cost of Sales Record.
  9. File Daily Sales and Receipts Summary form in permanent binder.

Purchases and Disbursements

1. Record all check stubs issued during the day to Purchase and Disbursements Record.
2. Carefully check accuracy of invoices as to price extensions and addition, and against shipment of merchandise received.
3. Record all invoices for purchases of merchandise to Purchases and Disbursements Record.
4. Post "Cash Paid Out" Section of the Daily Sales and Receipts Summary to the Purchases and Disbursements Record.
5. Post the Expense column entries to the Expense Distribution Record.
6. Post Accounts Payable and Notes Payable columns to the individual suppliers' accounts in the Accounts Payable and Notes Payable Ledgers.

Monthly Bookkeeping Time Table

Sales and Receipts

1. Total and balance the Sales and Receipts Record.
2. Summarize the several miscellaneous columns.
3. Post all columns to individual accounts in the General Ledger.

Purchases and Disbursements

1. Record accrued expenses (interest, wages, depreciation, etc.) and earnings for the month.
2. Examine bank statement and record any bank service charges or sight drafts in the Purchases and Disbursements Record.
3. Total the "Bank Withdrawals" column and balance the bank statement with the book figures.
4. Total and balance the Purchases and Disbursements Record.
5. Summarize the several miscellaneous columns.
6. Post columns to individual accounts in the General Ledger.

Cost of Sales

1. Determine Cost of Sales of Parts, and all other small ticket items by using a percentage of sales, and make entry on the Cost of Sales Record for each.
2. Total the columns of the Cost of Sales Record.
3. Post column totals (Except Misc. and Used Equipment) as DEBITS to the respective Inventory accounts. The Misc. and Used Equipment columns must be summarized, and the items posted as debits to Cost of Sales accounts and credits to inventory accounts as above.

General

1. Total the balances in the Accounts Payable Ledger and make certain that it agrees with the General Ledger Account, "Accounts Payable."
2. Total the balances in the Notes Payable ledger and make certain that it agrees with the General Ledger Account, "Notes Payable."
3. Total the balances in the Accounts Receivable (Customers Ledger) and make certain that it agrees with the General Ledger Account, "Accounts Receivable."
4. Total the balances in the Notes Receivable (Customers Ledger) and make certain that it agrees with the General Ledger Account, "Notes Receivable."
5. Prepare a Trial Balance of the General Ledger Accounts, to make sure that the total DEBITS equals the total CREDITS.
6. From the Trial Balance, prepare the Financial and Operating Statements.

Accounting terms and their everyday use quite often mean different things. Profit to the layman quite often turns out to be margin to the accountant. For this reason, it is very necessary for management to understand the language of the accountant. Here are some of the terms you should know.

Cost.   Amount of invoice from manufacturer or wholesaler, plus freight, assembly and handling expense. Freight is always a part of the cost. So also is the cost of preparing a machine to sell.

Overhead.   The cost of doing business, including all operating expenses, credit losses, and depreciation. As a percentage, overhead is the relation of the amount of the overhead to the volume of sales. In a business of $100,000 volume, with a cost of doing business of $16,000, the overhead is 16%.

Markup.   Amount added to cost to make selling price. As a percentage it is the ratio of the amount to the cost. If an article costs $75 and is marked up to sell for $100, the markup is $25, or 33% of cost.

Margin.   The difference between cost and selling price. As a percentage it is the ratio of the amount to the selling price. If an article costs $75 and is sold for $100, the margin is $25, or 25% of the selling price.

Markup and margin are often confused. They are always the same in dollars and cents but never the same in percentage, because markup is figured on cost and margin on selling price.

Turnover.   Turnover is computed when the cost of the goods sold during a fiscal period is divided by the average of the opening and closing inventories. This gives the number of times the average inventory has been "turned" during such a period. For example, if the cost of goods were $120,000 and the average of the opening and closing inventories were $20,000, the merchandise inventory turnover would be six times.

Profit.   Amount by which selling price exceeds cost and overhead. As a percentage it is the relation of amount to selling price. If selling price does not exceed cost and overhead, there is a loss. A sale on which there is a margin of 25%, by a dealer show overhead is 20%, yields a profit of 5%. A sale on which there is a markup of 25% by a dealer whose overhead is 20% means that he has just traded dollars without any profit whatever.

Handling Charge.   A charge by the seller for handling or servicing new equipment, such as erection and assembly, installation of all attachments, delivery of new equipment and removal of trade-ins, fuel and oil placed in the equipment, and training demonstrations.

Installation Charge.   A charge for special installation of equipment. It sometimes takes the place of the handling charge.

Base Price.   Usually the manufacturer's current published list price, f.o.b. factory, sometimes known as "suggested retail price."

Net Invoice Cost.   This is the invoice cost, less any discount or allowance. It does not include separately stated charges such as freight or taxes.

RESPONSIBILITIES OF MANAGEMENT

The dealer who is too busy to learn about his books does not last long in the business. Conversely, the dealer who feels he must make all the bookkeeping entries will not last long in the business. There is a happy compromise. Know the qualifications of a good bookkeeper, then hire one. Insist that the bookkeeper work for you. This means that analyses are computed in a timely manner, enabling decisions to be made so that they can affect the future course of events. Know what figures you will need and when. Establish timetables and standard operating procedures so that the recordkeeper can perform the job in a timely manner.

The dealer must know how to read financial statements. From this he should be able to uncover every possible area in which business improvements can be made.

In the financial analysis, net profit is probably the best test of management. Net profit is determined from (1) margins and (2) expenses. These two items should be the starting point of any analysis. These are usually expressed as a percentage of sales.

From the operating statement review your margin. Look at the margin for the overall business, this month, and year-to-date and compare to your goal for the year. Look at the margins by department—for the month and year-to-date—and compare to your goal for that department. Are you meeting or exceeding these goals? Which department has the problem? Why are margins lower than anticipated? Any of the following might be causing your problem:

Inaccurate physical inventory
Wrong price charged to the customer
Invoices entered incorrectly
Invoices entered on inventory but never received
Parts installed on shop equipment without being charged out
Lost repair orders
Lost sales tickets
Mathematical errors
Parts loaned out and not recorded
Pilfering
Unauthorized discounts
Improper pricing method

There are other possibilities as well. The preceding apply particularly to parts but may well serve as a guide for other departments as well. Remember that "the men do well what the boss checks on." Personnel will be more careful when they know you are concerned about margins.

The other half of the equation is expenses. Study them as a percentage of sales. Are your overall expenses as a percentage of sales less than the margin? Is it over or under your goal? Review department expenses. Are you allocating expenses to departments arbitrarily, or do your figures represent actual conditions? To the extent that expenses can be charged directly to the departments, your job of controlling expenses will be simpler. Management and administration will nearly always have to be arbitrarily assigned as a percentage.

Wages are a substantial portion of the total expenses. There is very little than can be done in this category beyond study in staffing guides. Review the Cost of Doing Business study. Compared to others with your volume of business, are you overstaffed? Where are you overstaffed?

Rent and utilities are also sizeable items. There isn't much that can be done about rent. Utilities can be wasted. Again, compare to past years and compare with other dealers of comparable volume.

Car and truck travel always need close monitoring. Will a telephone call do as well as a personal call? What about the telephone bill? Long-winded long-distance calls can really eat up profits. Prior planning can remove the need for many long-distance calls.

What about bad debts? Are they in line with expectations? Are they concentrated in one department? Could a collection agency improve this area?

The miscellaneous column is always suspect. Is it a means of hiding unnecessary items?

Operating expenses must be controlled. It isn't always possible or even desirable to reduce expenses. Additional money spent on advertising may improve profits. Additional money spent as interest to finance receivables may be money well spent. There is really only one criterion to use. Are these expenses resulting in additional sales and profits? If they are, the expense in dollars will rise but not the expense as a percentage of sales.

Net profit is determined by margins and expenses, but since these are expressed as a percentage of sales, we cannot overlook total sales. Perhaps this may best be illustrated by an example. A business with a net worth of $160,000 has sales of $1,000,000, $170,000 margins (17%), and $140,000 expenses (14%). This results in a net profit of $30,000 (3%). A net profit of $30,000 on a net worth of $160,000 yields a return on net worth of 18.75%.

If we increase the margin to 18% and hold sales and expenses constant, we increase net profit to 4% or $40,000. This will yield a return on net worth of 25%.

If we decrease expenses 5% $(0.05)($140,000) = ($7000)$, we now have a net profit of $170,000 − $133,000 or $37,000. This yields a net profit of 3.7% and a return on net worth of 23.125%.

Both of these changes improved net profits and return on net worth. If we combine these with an increase in sales, the effect is dramatic:

Increase in sales (3%):
Total sales: $1,030,000
Increase margins (3%) to (17.51):

New margin: $180,353

Decrease expenses (3%) to (13.58)

New expenses: $139,874

Net profit $40,479
Return on net worth 25.29%

So you're maintaining margins, controlling expenses, selling effectively, and turning out a healthy net profit. What else do you look for?

Study the balance sheet. Of particular importance is your cash position. Are you in a position to take advantage of cash discounts? Can you weather the next two months without having to resort to distressed sales of inventory? High-profit dealers are maintaining about 7% of their assets in cash and securities. Low-profit dealers have closer to 2.5% in cash and securities.

Look at the receivables next. How does it compare to your goal? Though not found in the balance sheet, study your aged accounts receivables. Are your receivables current? Anything over 90 days should be a cause for concern. Are your receivables in proportion to your total sales? How do they compare with the last two or three years?

Inventory is always cause for concern. It is accurate? Does it reflect realistic values? Look carefully at your used inventory. When it exceeds 15% of your total inventory, become concerned. Look at inventory turnover. Are you meeting goals?

Current liabilities are a necessary evil. Are they manageable? Current assets minus current liabilities equals working capital. Do you have sufficient capital? What is your capital turnover? Perhaps you have too much capital invested.

Finally, look at your net worth. It is to be hoped that it is steadily getting bigger and bigger.

PRICING MERCHANDISE

Equipment dealers can stay in business and render a service to a community only as long as they are able to sell at a profit. The manager must see that his merchandise is priced competitively but profitably. The selling price must cover cost of sales, overhead, and a reasonable profit. This requires information that your bookkeeper should provide. When pricing an item, first look at cost of sale. Remember, it is made up of:

Invoice cost
Plus freight
Plus assembly, etc.
Plus any taxes not passed directly to the consumer

Next, add overhead. This figure should be obtained from the department making the sale. No sale should fail to exceed this total amount. Consider the following example:

Manufacturer's invoice cost	$3000.00
Freight	300.00
Assembly and handling	150.00
Overhead expense	300.00
Total cost	$3750.00

Knowing this total, it is evident to the dealer what his minumum sale price can be. Only by knowing current figures by departments can management set competitive but fair prices for merchandise.

Generally, it is preferable to price according to a certain margin. That margin should be set as a percentage of sales that will exceed overhead by the percentage net profit desired. Suppose your overhead in whole goods is 8% of sales, and your desired profit in whole goods is 3% of sales. A margin of 11% of sales is set. The selling price therefore becomes:

$$\text{Selling price} = \frac{\text{Cost of Sale}}{1 - \%\,\text{Margin}}$$

As an example, consider an item with a cost of sale of $3450 and a desired margin of 11%.

$$\text{Selling price} = \frac{\text{Cost of sale}}{1 - \%\,\text{Margin}}$$

$$\text{Selling price} = \frac{\$3450}{1 - 0.11} = \frac{\$3450}{0.89} = \$3876.40$$

If the overhead is 8%, the total cost of the item is

$3450.00	Cost of Sale
+ 310.11	Overhead ($3876.40 × 0.08)
$3760.11	Total Cost

$3876.40 − $3760.11 = $116.29 net profit (3% of $3876.40)

Margins in different departments will be quite different. In the parts department, a margin of 30−40% is quite common. Used equipment on the other hand may at times command a much lower margin.

Markup is often confused with margin. Markup is based on cost. Margin as was pointed out previously, is based on selling price. In dollars, the amount is the same. In percentages, they are quite different. Some dealers prefer to price using markup rather than margin. How do we determine what markup will yield a certain margin?

$$\text{Markup} = \frac{\text{Margin } (\%)}{(1 - \text{Margin } \%)}$$

where: Margin = Percentage of selling price
        Markup = Percentage of cost

As an example, let us consider the previous case where the cost of sale was \$3450.00 and a margin of 11% was desired. What should the markup be?

$$\text{Markup} = \frac{\text{Margin}}{(1 - \text{Margin})} = \frac{0.11}{(1 - 0.11)} = \frac{0.11}{0.89} = 0.1236 = 12.36\%$$

Markup multiplied by cost of sale yields markup in dollars. Markup (dollars) plus cost of sale equals selling price.

\$3450	Cost of sale
× 0.1236	Markup (percentage)
\$426.42	Markup (dollars)
+ \$3450.00	Cost of sale
\$3876.42	Sellling price

Either pricing method is acceptable as long as you know your cost of sale and your markup in dollars or your margin in dollars is greater than your general overhead in dollars. This will yield a net profit. Table 8.1 shows what markup on cost of sales is necessary to yield a given gross margin.

TABLE 8.1. MARKUP REQUIRED TO YIELD GIVEN MARGIN OF SALES

Margin of sales (%)	Markup of cost of sales (%)	Margin of sales (%)	Markup of cost of sales (%)
10	11.11	24	31.57
11	12.36	25	33.33
12	13.63	26	35.13
13	14.94	27	36.98
14	16.27	28	38.88
15	17.64	29	40.84
16	19.04	30	42.85
17	20.48	32	47.05
18	21.95	34	51.15
19	23.45	36	56.25
20	25.00	38	61.29
21	26.58	40	66.66
22	28.20	45	81.81
23	29.87	50	100.00

INVENTORY RECORDS

The bookkeeper is also the card keeper. One of the more important cards is the inventory card. These cards apply to parts and new and used inventory. They may be kept manually or they may be processed electronically.

The problem of inventory values of trade-ins is important enough to deserve discussion here. Whole goods are normally received with a posted list price, which is a manufacturer's suggested sale price. Competition may dicate selling at some price below this. When trade-ins are involved, the new machinery is sold at list price and the customer is given an inflated price for his used machinery. This is an accepted practice, but it can lead to erroneous financial statements. This happens if the trade-in is picked up on inventory at the inflated price.

There are two methods of handling this transaction which will give accurate financial information.

Consider this example. A sale is made for $9000. The suggested list price of this equipment is $10,000. A 10% discount was given off the list price. The actual invoice cost of the new equipment was $7700. Used equipment was taken in trade with an as-is value of $3000. With these facts, the transaction can be handled as follows:

Method 1.   The new equipment is listed as a sale of $9000.When the invoice cost is deducted, the actual margin shown is $1300. The used equipment is put into the used equipment inventory at a value of $3000. This is the fair market value. This method will show absolute accuracy on the financial statement.

Method 2.   Some dealers refuse to make a sales ticket for less than the list price. To do this, the used equipment is over-valued. When this is desirable, an over-allowance account should be set up. The used equipment shown on the sales ticket at $4000 should be re-valued and placed in inventory at $3000. The extra $1000 is then carried in the over-allowance account as a credit balance. The inventory and profit are now correctly stated.

It is obvious that if the used equipment is put in inventory at the inflated value, a false profit will be shown if the equipment is not sold before the end of the month. Several transactions like this can really give an unsuspecting dealer a false sense of accomplishment.

Inventory Cards Serve a Dual Purpose

Typical of inventory card records is the Machine Inventory and Sales Record card distributed by several state associations (Fig. 8.2). This provides the dealer with an inventory record of all whole goods on hand. It tells him what is in the warehouse or being repaired and reconditioned, without the necessity for trips to the warehouse, which may be some distance from the store or salesroom.

The most important use for the inventory card is that of providing a quick and easy means of determining the cost of each item sold. Present-day accounting procedures insist that every farm and power equipment dealer should know the exact margin he receives on all sales of new and used equipment. Such knowledge takes all guesswork out of recordkeeping, a procedure which is of untold value when operating in present day highly competitive markets. These inventory cards provide ample space for recording the actual cost of each piece of equipment that goes into inventory. When this actual cost is deducted from the actual selling price, the dealer has actual margin as the remainder. When such information is combined from all sales, the total margin revealed on the profit-and-loss statement is a true picture of his ability to sell his products at a profit. No longer must the dealer wait until an inventory period arrives to determine his true or actual margin on the sales he has made. With the inventory card incorporated in his recordkeeping process, he can look at his operating statement with assurance that it reveals a true statement of fact.

An added use for these cards, after they have served as inventory and cost accounting records, is for after-sale records. They are usually filed alphabetically according to purchaser's name, once the sale has been completed.

 # MACHINE INVENTORY & SALES RECORD

KIND OF MACHINE		LOCATION		SALES DATA	
MAKE	MODEL	SIZE OR CAPACITY		SOLD TO:	
SERIAL NO.	MOTOR NO.	TYPE		ADDRESS	
DATE PURCHASED	PURCHASED FROM	INVOICE NO.			

RECORD OF INDEBTEDNESS	INVOICE COST, NEW	$	SALES TICKET NO.	DATE
NOTE NO.	EXCISE TAX			
DATE	FACTORY FRT.		LIST PRICE	$
DUE DATES	WILL CALL OR TRUCKING EXPENSE		ATTACHMENTS:	
	ASSEMBLING AND HANDLING:			
BALANCE	REPAIR ORDER NO.			
ORIGINAL AMT. $	REPAIR ORDER NO.			
DATE PAID_____	REPAIR ORDER NO.			$
PRIN._____	ATTACHMENTS:			
INT._____ $			FREIGHT	
DATE PAID_____				
PRIN._____			HANDLING	
INT._____ $			TOTAL	
DATE PAID_____				
PRIN._____			TRADE-INS	
INT. $			TAXES	
INVENTORY COST				
ACCT._____ $	TIRES		BALANCE SETTLEMENT $	
ACCT._____ $	TOTAL COST	$	CASH: $_____ NOTE: $_____ ACCTS. REC. $_____	

SALES AND MARGINS		REPAIR RECORD			
GROSS SELLING PRICE	$	DATE	R. O. NO. AND DESCRIPTION		AMOUNT
LESS OVERALLOWANCES: TRADE-IN NO._____	$_____				
TRADE-IN NO._____					
TRADE-IN NO._____	$				
SUB TOTAL	$				
FREIGHT, TAX, HANDLING, ETC.					
TOTAL SALES					
LESS COST					
GROSS PROFIT					
GAIN (OR LOSS) ON TRADE-INS:					
GROSS PROFIT AFTER SALE OF ALL TRADE-INS	$				

FIG. 8.2.   Machine Inventory and Sales Record card, front (top) and back (bottom) views.

To illustrate their use, when tractors are received, either new or used, a card is made for each machine. Each card describes the machine and its standard attachments. The invoice cost and freight and handling charges are also recorded. The card is then filed in the inventory control box under tractors, new or used, as the case may be. While in inventory, the card affords ready reference, indicating exactly what machines are in stock and pertinent information necessary for selling.

When a tractor is sold, the appropriate card is lifted from the file and the "cost" is recorded simultaneously with the selling price. The name and address of the purchaser is also recorded on the card along with the selling price. When completed, the card is filed alphabetically according to purchaser's name and thus becomes a permanent record. Some customers feel a dealer should be like an elephant and never forget. They expect him to remember the exact set-up, attachments and all, for as many years as the tractor or equipment may remain in service. Admittedly, this may become confusing. But with the permanent record to fall back on, the whole detail can be brought to light at a moment's notice. In case the serial number has been forgotten, a glance at the card will give the information needed so that exactly the right parts can be made available for service.

Another use for the card is as an aid in making after-sale service calls. When dates of service calls are recorded and the extent of the call is known, a dealer can be guided on his policy for additional service that may be needed. These cards also provide an excellent reference from which future sales may be made.

In this manner, the bookkeeper or recordkeeper provides a means whereby a dealer can better serve his customer and at the same time reduce his expenditure of time and money. It is a simple record to maintain and can be obtained from any state or regional association office.

PROSPECT RECORDS

The prospect record serves as a management tool for the sales manager and as a valuable sales ally to the dealership. It is an effective sales tool.

Prospects are acquired in personal contacts on the farm, through visits of customers to the dealership, through mail advertising, and through sign-in rosters at various open house events. Regardless of how the information is acquired, once a contact is made and interest in a particular piece of equipment is evidenced, the salesman should fill out a prospect card. Typically, it will list his name, date, address, telephone number, prospect for ———, result of contact, and date for return call. The card is then filed.

At some later date, perhaps two weeks, a return call is made. The salesman points out the advantages of this equipment, gets agreement on desirable options, and quotes a price. The prospect is not quite sold yet, but his card is flagged. This means another call should be made soon. A third call may be made again, perhaps a week later. Perhaps the customer comes into the store while the salesman is out. The "boss" takes over. Looking at the prospect card, he knows what has been quoted and how much. He quotes exactly the same. Plans are made for the salesman to appraise his trade-in. The next day, order pad in hand, the salesman returns to the customer and confirms the equipment needed and price. He appraises the trade-in and completes the transaction. The prospect card has played a vital role in this sale.

In addition to personal calls, prospects are acquired in many ways. Train your parts man to recognize when a customer is ready to stop buying parts and start buying new. The service manager is also in an admirable position to find out who needs what. When sending out monthly statement, include literature about equipment you think a customer may be interested in. Alert personnel in a dealership should recognize and record specific interest of visitors to the store. Particular curiosity about equipment is reason to pursue the reason for interest. When holding an open house or other public relations function, have a sign-in roster. Never miss a chance to find out what the potential customer needs. Remember, you have first to catch a rabbit before you can make a rabbit stew. The prospect card is often the trap needed.

THE PROFIT CONTROL PLAN

No record is more important than your "Profit Control Plan." Yes, the successful businessman does control profit. Chart a path to profit if you want it to happen. That is the basic purpose of the plan. It points the direction and establishes guide posts and check points all along the way.

Each dealership should establish goals. Your profit control plan should incorporate most of those goals into the plan. Some management goals will not appear directly in the plan but actually still play a vital part in the profitability of the dealership. The goals become your milepost toward your profit objective.

There are several variations of the profit control plan. Since the exact format is not important let's consider the plan made popular by the National Farm and Power Equipment Dealers Association. It uses five work sheets, shown in Fig. 8.3. If the plan is to be successful, it has to be made up by all the major departments. Avoid the temptation of letting the bookkeeper or the general manager make the plan. They certainly must finalize the plan, but without the concurrence of the department heads, the plan is doomed to failure.

Work Sheet No. 1 is the expense estimate. That sheet includes a line for each expense classification and column for the past three years plus a column for the estimate for the coming year. The bookkeeper can fill the first three columns. The estimate for the coming year requires the best thinking of all the department heads. A word of caution here is in order. You cannot change your expenses a large amount very quickly. Salaries and employee benefits are about 48% of your expenses. You can expect those categories to go up. You can reduce this expense, but only as a percentage of sales and only if you can make your labor more productive. There are certainly some economies that can be made in all businesses. The easiest and most productive way to decrease expenses is to increase sales productivity. Remember that expenses are important mainly as a percentage of sales.

FIG. 8.3.   Profit control plan work sheets. A. Expense estimate work sheet. B. Estimated Profits and Margins, Estimated Total Sales, and Estimated Departmental Sales and Margins work sheets. C & D. Estimated monthly goals, sales and margins work sheet.

WORK SHEET NO. 1 - EXPENSE ESTIMATE

F.E.R. 500-A Acct. No.	EXPENSE CLASSIFICATION	Three Years Ago	Two Years Ago	Last Year	Estimate This Year	Revised Estimate
31	Salaries, Officers and Owners					
32	Salaries, Office Employees					
33	Salaries, Salesmen					
34	Salaries, Partsmen					
35	Salaries, Other Employees					
36	Rent & Lease Expense					
37	Heat, Light, Power & Water					
38	Telephone & Telegraph					
39	Postage					
40	Store Supplies & Equipment Repairs					
41	Office Supplies & Equipment Repairs					
42	Shop Supplies & Equipment Repairs					
43	Car & Truck Expense					
44	Gas & Oil					
45	Travel & Sales					
46	Advertising					
47	Payroll Tax, F.O.A.B.					
48	Payroll Tax, Unemployment					
49	Taxes — Real Estate and Property					
51	Other Taxes & Licenses					
52	Dues & Subscriptions					
53	Insurance (Except Group)					
54	Legal & Auditing					
55	Demonstrations					
56	Bank Charges					
57	Group Insurance					
59	Repairs, Bldg. & Bldg. Fixtures					
60	Depreciation					
61	Bad Debts					
64	Interest on Notes Payable					
65	Interest on Mortgages					
67	Donations					
68	After Sale Expense					
71	Miscellaneous					
	COST OF DOING BUSINESS (Total of above)					

WORK SHEET NO. 2 — ESTIMATED PROFITS & MARGINS

LINE		1st ESTIMATE	REVISED EST.
1	Desired Profit	$	$
2	Add: Income Tax Estimate		
3	Desired Profit Before Taxes		
4	Add: Estimated Expenses (Total of Work Sheet No. 1)		
5	Desired Total Margin (Enter in Worksheet No. 3, Line 4)		

WORK SHEET NO. 3 — ESTIMATED TOTAL SALES

LINE		1 3 YEARS AGO	2 2 YEARS AGO	3 LAST YEAR	4 ESTIMATE THIS YEAR	5 REVISED ESTIMATE
1	Total Sales	$	$	$	XXX	XXX
2	Margin	$	$	$	XXX	XXX
3	Margin % of Sales	_______%	_______%	_______%	_______%	_______%
4	Margin Desired	XXX	XXX	XXX	$	$
5	TOTAL SALES REQUIRED (Divide Line 4 by Line 3)	XXX	XXX	XXX	$	$

WORK SHEET NO. 4 — ESTIMATED DEPARTMENTAL SALES & MARGINS

LINE		1	2	3	4		5
	SALES BY DEPARTMENTS	3 YEARS AGO	2 YEARS AGO	LAST YEAR	ESTIMATE THIS YEAR		
1	New Farm Eqpt.						
2	Used Farm Eqpt.						
3	Repair Parts						
4	Service Labor						
5							
6	TOTAL SALES						
	MARGINS BY DEPARTMENTS	3 YEARS AGO	2 YEARS AGO	LAST YEAR	ESTIMATE THIS YEAR	% SALES	% SALES
7	New Farm Eqpt.						
8	Used Farm Eqpt.						
9	Repair Parts						
10	Service Labor						
11							
12	TOTAL MARGINS						

WORK SHEET NO. 5—ESTIMATED MONTHLY GOALS—SALES & MARGINS

NEW FARM EQUIPMENT

MONTH	1 3 YRS. AGO	2 2 YRS. AGO	3 LAST YEAR	4 EST. THIS YR.	5	% Sales (From Work Sheet No. 4, Line 7.)	6 ESTIMATED MARGIN
JAN.							
FEB.							
MAR.							
APR.							
MAY							
JUNE							
JULY							
AUG.							
SEPT.							
OCT.							
NOV.							
DEC.							
TOTAL							

USED FARM EQUIPMENT

MONTH	1	2	3	4	5	% Sales (From Work Sheet No. 4, Line 8.)	6
JAN.							
FEB.							
MAR.							
APR.							
MAY							
JUNE							
JULY							
AUG.							
SEPT.							
OCT.							
NOV.							
DEC.							
TOTAL							

REPAIR PARTS

MONTH	1	2	3	4	5	% Sales (From Work Sheet No. 4, Line 9.)	6
JAN.							
FEB.							
MAR.							
APR.							
MAY							
JUNE							
JULY							
AUG.							
SEPT.							
OCT.							
NOV.							
DEC.							
TOTAL							

WORK SHEET NO. 5 — CONT'D.

SERVICE LABOR

MONTH	1 3 YRS. AGO	2 2 YRS. AGO	3 LAST YEAR	4 EST. THIS YR.	5	% Sales (From Work Sheet No. 4, Line 10.)	6 ESTIMATED MARGIN
JAN.							
FEB.							
MAR.							
APR.							
MAY							
JUNE							
JULY							
AUG.							
SEPT.							
OCT.							
NOV.							
DEC.							
TOTAL							

MONTH	3 YRS. AGO	2 YRS. AGO	LAST YEAR	EST. THIS YR.		% Sales (From Work Sheet No. 4, Line 11.)	ESTIMATED MARGIN
JAN.							
FEB.							
MAR.							
APR.							
MAY							
JUNE							
JULY							
AUG.							
SEPT.							
OCT.							
NOV.							
DEC.							
TOTAL							

MONTH	3 YRS. AGO	2 YRS. AGO	LAST YEAR	EST. THIS YR.		% Sales	ESTIMATED MARGIN
JAN.							
FEB.							
MAR.							
APR.							
MAY							
JUNE							
JULY							
AUG.							
SEPT.							
OCT.							
NOV.							
DEC.							
TOTAL							

In Work Sheet No. 2, we set down our desired profit, add our estimated expenses from Work Sheet No. 1, and determine the needed dollar margin to yield that profit.

Work Sheet No. 3 is the sales estimate. Start by recording the sales and margins of the past three years. Now you must estimate margins as a percent of sales for the coming year. Again, don't expect to change radically the margins you have been maintaining. It takes good salesmanship to improve margins. Since this figure is a composite figure for the dealership, it is important that each department knows what margin it is expected to generate. Remember, too, that margins will vary widely by departments.

A mix of 12% for new whole goods, 8% for used goods, 30% for parts, and 35% for labor will give an overall margin of 15.2% of sales. This is based on sales of 59% whole goods, 20% used, 15% parts, and 6% labor.

Divide the total margin desired, Line 4, Work Sheet No. 3, by the percentage margin, Line 3, and you will now have your dealership sales goal.

$$\text{Sales} \times \text{percentage margin} = \text{margin in dollars}$$

$$\text{Sales} = \frac{\text{Margins (dollars)}}{\text{Margins (\% of Sales)}} = \frac{\text{Line 4}}{\text{Line 3}}$$

Work Sheet No. 4 divides the company goal into departmental sales and margins. Many of you will find that it is more realistic to fill out Work Sheet No. 4 before you do Work Sheet No. 3. After all, the company can only do what the departments do.

Work Sheet No. 5 further divides all the departmental sales and margins goals by months. Don't forget this last but very important step. This is what you use to make up your monthly charts. Post them conspicuously where all can see them.

Sounds like a lot of work? Perhaps, but not near as hard or disagreeable as trying to borrow money to operate on when your operating statement shows a loss. It need not be as bad as it seems. Once the projections are made, let the computer keep up with the monthly progress. Your "Operating Trends and Budget Report" will show you your variance by categories. Dig into those categories that are lagging behind. Find out why, then do something about it.

Examples of the computerized version of the profit control plan are shown in Appendix C. The "Budget Projection Worksheet" is used to initiate budgets. The "Sales and Margins" projection shows the completed estimates of sales, margins, and expenses for each month of the year by quarters. The "Operating Trends and Budget Report" shows how well you are meeting your sales goals. The "Expense Trends and Budget Report" shows how well you are controlling expenses. The "Operating Trends—Three Years" is also available to enable you to better evaluate long-range trends.

An Alternate Method.   Let the departments start by completing Work Sheet No. 5. This is their goal—sales and margins by the month. Use Work Sheet No. 4 to summarize their goals into a company goal. Collectively fill out Work Sheet No. 1, the Expense Estimate. Where expenses are broken down by departments, each department should make his own; then a company estimate can be completed as Work Sheet No. 1. Next complete Work Sheet No. 3. Total

Sales would then be obtained from Work Sheet No. 4 rather than computed as indicated on Work Sheet No. 3.

Finally, a new summary work sheet might be used, as follows:

Summary Work Sheet

Line	1st Estimate	Revised
1. Anticipated Margin	________	________
2. Anticipated Expenses	________	________
3. Profit before Taxes	________	________
4. Income Taxes	________	________
5. Net Profit	________	________

If the net profit resulting is not satisfactory, then there are four alternatives:

1. Increase projected sales
2. Increase projected margins
3. Decrease expenses
4. Combinations of 1, 2, and 3

The alternate method has the advantage of giving greater participation to the departments. Control still rests with the manager, for he has final approval of the plan. The column in all the work sheets for revised estimates is used to update the plan any time circumstances require it. Conditions may improve or degenerate during a one-year period.

OFFICE PROCEDURES

Small businessmen in general tend to neglect the office and its procedures. It is a necessary part of the business. It deserves the attention of management. Without a successful administrative routine, no business can long survive.

First and foremost is a private area free from the distractions of the operating sections of the business. It need not be fancy but should be comfortable and well lighted. An area in a throughway just won't do.

Equipment needs are quite variable. They may range from a simple desk, chair, filing cabinet, and typewriter to multiple desk offices with electronic data processing equipment, billing machines, and so on. Regardless of its complexity, enough equipment must be provided to supplement the manpower needs or to replace them in part.

Systematic filing is essential in every business. There should be a place for everything and everything should be in its place. Business supply houses can be helpful with this task.

Important Procedures

Earl Thompson, Director of Services of the Retail Farm Equipment Association of Minnesota and South Dakota, had these recommendations for local dealers:

Filing System Important

The filing system should be systematic and adequate. A to Z dividers can be used to advantage in certain sections of the files such as the part for paid and unpaid invoices. Several types of file folders with name and tabs are available.

A special area should be reserved for invoices and statements from suppliers. This area should be divided into the paid section and the unpaid section. With good office management, a dated file will probably do for unpaid invoices. The discount date on the invoice or statement determines the place where such are filed. Once payment has been made, statement and invoices are then filed in the "paid" section according to company name. Any supplier with whom a large volume of business is transacted will require at least one folder and frequently several are used to advantage.

Wherever possible a good inventory card system for new and used equipment should be maintained. This eliminates the necessity for employees constantly referring to invoices to find the cost. Every association can supply a dealer with inventory cards that will prove valuable.

Don't Forget the Customer File

A customer file will be extremely helpful as a source of information. It should be a section containing a separate folder for each customer with whom there has been dealings or correspondence. Orders, claims, correspondence, service records, personal data, credit records, human interest items . . . all of these and much other information can be preserved in the customer's file. Whenever information is needed about the history of an account, it can then be found under the customer's name. The only exception would be such records as sales tickets, shop tickets, prospect records and after-sale records which more logically may be kept in a numerical file or a separate alphabetical file.

The Miscellaneous Section

Every dealership has considerable miscellaneous material which must be preserved such as records on insurance, taxes, advertising, correspondence from other dealers, association material and a considerable quantity of other material which needs to be preserved for future reference. A section of the file should be devoted to such materials. The very important items such as insurance policies, contracts, leases, etc., will be more secure if stored in a fireproof safe.

Suppliers' material should be separated into several files such as the invoices section previously discussed, correspondence, promotional material, claims, etc. Combinations of material can be made in cases where the quantity from a given source is limited. It is a good policy to attach invoices and remittance sheets to monthly statements and file the combination under paid bills.

Promotional material as well as operational data is important to the business and will require some thought as to the best method of handling. Much of it can be used in display racks but there will be some information which is best preserved in the files. It can be filed in either of two ways . . . by company name . . . by kind of machine.

Routine Procedure Necessary

Most of the papers handled will arrive through the mails. Before anything can be accomplished the MAIL HAS TO BE OPENED. This is a simple task, yet it seems to be one of the more difficult ones for dealers to accomplish. Some of the mail may fail to get opened and will accumulate in great piles on the dealer's desk.

Opening the mail is a routine job and should be treated as such. There shouldn't have to be a major decision made every day about getting it done. Give someone in the organization the responsibility for getting it done every day at a certain time. In the smallest dealerships where clerical help is not employed, the owner may find that opening the mail is one of his duties. But whoever opens it, the fact remains that opening the mail is a duty that must never be neglected. By making it a routine job, one will never be tempted to put off the problem by ignoring it.

Although the mail is opened, it can still be stacked and piled without being read by anyone. The dealership where this occurs will usually have very little room on a desk that is piled high with several months' accumulation. How can one best eliminate the stacks of material?

By starting as the envelopes are opened! As the contents are removed, make sure you have all of the needed information before destroying the envelope. It's possible a return address may be needed. The envelopes may then be put in the wastebasket. Next, sort by person to receive the item, or by filing section.

As an example, one might designate one pile as throwaway. There is some mail every day that has no interest of value. Information material may be put in another pile. Invoices, statements, correspondence, cash receipts, magazines and other miscellaneous material may be sorted and classified.

If the business is large enough to employ office help, the owner or manager will be relieved of the job of opening the mail but should want to look it over before it is distributed. Machine invoices, correspondence, and most other mail is important enough to be scanned by the manager.

With good office help, it is possible for the manager to give most of the routine paperwork to another employee, but the one job he will always have to do himself is take care of the correspondence. Some of it requires that a decision be made before writing the letter.

Answer Corrrespondence Promptly

Here we come to the same old problem of putting off until tomorrow. Answer correspondence immediately, if possible. How it is answered is not as important as the fact that it is answered. Longhand, typewriter, dictation to secretary, post card, letter, memo—use the best method available but take care of it now.

Keep on Top of the Mail

For the dealer without full-time office help, a dictaphone may be a good solution. He can then dictate at his convenience. Whatever method is used in writing letters, always make a carbon copy. This is proof positive of what was said in the letter.

A come-up file is a real help when letters can't be answered immediately or there is need to check on your correspondence for an answer. Normally one file folder will be sufficient to hold everything.

When it is necessary to receive an answer, make an extra copy of the letter and file it with a date notation. This can be done on the letter or by attached slip. File the copies by date with oldest dates in front. Each day or two go through the file and find out which answers are due. A follow-up letter with another come-up date will probably be necessary.

If a letter is received asking for information that is not readily available, write to the party and explain. Put the letter with a copy of the answer in the come-up file with a date notation. A daily check of the file will turn up the letter and carbon on the proper date.

Summing up a few important rules to be followed to get the paperwork done:
Providing good working condition.
Open the mail.
Make a daily routine of all paperwork.
Follow the routine—"Do it Now."
Have a place to file every piece of paper.
Each employee has definite responsibilities.
Method is not as important as the fact that the job gets done.
Check with your Association on available forms and assistance.

Set up a system covering the entire process from the time you open the mail until final disposition is made of every piece of paper, especially if more than one person uses the files. Your own special filing system may be a little hard for others to understand.

QUESTIONS

1. What is a balance sheet?
2. What information can be obtained from the profit-and-loss statement?
3. How do you compute turnover?
4. What two bookkeeping items determine net profit?
5. What constitutes cost of sale?
6. What is the difference between mark-up and margin?
7. Explain how you would handle an over-allowance in order not to show a false profit?
8. Of what value is a profit control plan?
9. Complete the accounting problem found in Appendix D.

General Management of the Business

Management is the unifying force that coordinates resources and directs action. The manager must have knowledge of business practices and must make decisions. Knowing what needs to be done and doing it are often two entirely different things.

The principles of management are explained in many texts. Though listed in many forms, they come down to four basic principles:

1. Analyze
2. Organize
3. Deputize
4. Supervise

Management is complicated by the fact that it exists at many levels. Management for the general manager is certainly different from management as practiced by the parts manager. As the level of management goes down, the amount of supervision involved goes up. The basic principles still apply.

TOP ECHELON MANAGEMENT

The top management of any firm has four specific duties for which it must take full responsibility, according to George Scherovich, editor of *Implement and Tractor*. The day-to-day operation can be assigned to others but the following four cannot be readily delegated:

1. Defining the purpose and mission of the business. The first thing that comes to mind is to make a profit. Profit is a good measure of management. It is necessary for the survival of any business. It is the consequence of management but not necessarily the purpose of the business. If this is the only purpose of the business, then it should be stated. It certainly would affect how business is conducted.

Most dealerships have the mission of providing farmers with production equipment and supplies to maximize their income. Whether or not counsel on agricultural production problems will be a part of the operation must be determined. The mission of the dealership must be firmly established and kept within bounds so that it can be financed.

2. Setting the goals. Set reasonable, attainable, and measurable goals. Set them in consultation with department managers. Goals are used to guide the business on its way to accomplishing its mission. Goal setting is a continuing task. As conditions change, goals must often be reevaluated.

3. Develop the organization. Top management must accept the responsibility for building the core organization. After deciding on the basic structure of the organization, lines of authority and responsibility must be laid out. Key department managers must be hired and oriented as to the policies of management. Clear-cut job descriptions for these key people should be carefully drawn up. This goes far toward enabling people to develop to their potential on the job.

4. Review and audit. Periodically and systematically, management must appraise the accomplishments of the organization. Are goals being met? Does the form of organization need to be changed? The principle of "management by exception" is applied here. Pay particular attention to goals that are being exceeded or not met.

SETTING GOALS

There are two popular styles of management being practiced today, "Management by Objectives" and "Management by Exception." In management by objectives, goals are set and efforts are directed toward reaching all of these goals. In management by exception, goals are set and effort is directed to those areas where goals are not being met or are being exceeded. The farm equipment sales store lends itself to the latter method since the manager of the typical dealership never seems to have enough time to monitor all the activities. Both methods are effective, but both require that goals be set.

Set reasonable, attainable goals. Set them in consultation with department managers. Combine these goals into overall management goals. Use them as checkpoints to gauge your progress. For this to be practical, goals have to be matched to a timetable. Goal setting is a continuing task. As conditions change, goals must be reevaluated.

Attainable and Measurable Goals

Let us consider some overall company goals. Growth is always one of the first to come to mind. It is easy to measure, but in determining how reasonable and attainable it is, be sure to consider inflation. Remember first to increase the expected sales by the inflation rate, then to add a suitable percentage for growth. Stated mathematically it is:

$$\text{Growth Goal} = [\text{Previous Sales} \times (1 + \text{Inflation rate})] \times (\text{Growth rate} + 1)$$

Dollar volume of sales is not a meaningful measure unless a specific margin is specified. Many will argue, and rightfully so, that margins without volume will soon take you to the poor house. The obvious answer is to set a dollar volume goal at a specified minimum margin.

Management has also to consider goals that cannot be expressed in dollars but do result in increased dollars. Inventory turnover is a most important one. With interest rates bordering on 20% we see immediately that a turnover of one will cost the dealer or the manufacturer and ultimately the consumer 20% of sales. A turnover of two will halve that cost, but a turnover of four only reduces it to five %. A reasonable, attainable ratio is probably 2.5. The salesman wants an unlimited inventory to sell from, but management must demand a reasonable turnover.

Management must further insist on goals that control expenses. The key word here is *control*, not reduce. Control may imply increased expenses in categories that will increase volume and ultimately reduce expenses as a percentage of sale. Control certainly means reducing expenses that are unnecessary and fail to contribute to the profitability of the business.

The nice thing about goals is that they apply to all departments. The service department may have quite different goals. Here, labor recovery rate should have a high priority. Is your shop selling at least 65% of its labor? Customer rates have to rise materially as recovery rates go down. Again, recovery rate is a goal that can be passed down to each individual mechanic.

Monitoring Goals

So you, in conjunction with your department managers, have set reasonable, attainable, and measurable goals. Remember their purpose. They are both the speedometer and the odometer that will indicate how fast you are traveling and how far you have gone. Keep up with the progress by posting, at least monthly, the contributions of all the departments toward the company goal. Build up some friendly competition between departments. You can plot the results on charts, graphs, thermometers, or speedometers. The exact form you use is unimportant. What is important is that all in your organization feel that their contributions are a vital part of the company goal. Let everyone see these charts and graphs if you expect your employees to feel that their efforts are most important in reaching the company goal.

QUALITIES OF A FARM AND POWER EQUIPMENT MANAGER

What does it take to be a good retail equipment manager? There are many necessary qualities, but the following are considered most important:

1.   Business sense.   This is an intangible quality called by various names. It's the fifth sense that makes you do the right thing at the right time. Don't confuse it with "hunches" and "premonitions." It's an ingrained sense of direction. It's an attitude that causes you to take advantage of a business opportunity. It's the quality of leadership that gives "espirit" and "morale" to an organization. It radiates confidence and can-do spirit to an organization. In summary, it's the sum of all the knowledge of a business, which perceptive people acquire and learn to use.

2.   Energy.   A good manager is never sluggish. It takes good health and a feeling of well-being to maintain an active interest in all phases of the business.

3.   Accounting.   The manager must be able to read and interpret a financial statement. He need not know all the details, but he must know where and how data is accumulated and recorded. He must have an appreciation for the financial controls the accountant develops. He must understand the problems involved, the time frame requirements, and the pitfalls that follow poor accounting practices.

4.   Capital.   The manager must know and have the capital requirements for inventory and operations. He must be able to develop ratios to guide him in the operation of the business. He must know how to manage money and keep accounts in balance.

5.   Finance.   The dealer must develop sources of credit for himself and his customers. It not only assists the customer, but provides revenue for the business as well. He must locate sources of credit and develop and maintain harmonious relations with them during the year.

6.   Credit.   Managing credit wisely can mean the difference between success and failure. Establishing a wise credit policy, good control, and an active collection procedure is an important part of the manager's job. Credit must boost not only sales but profits.

7.   Product knowledge.   The good manager must know his products and be personally convinced of their superiority. His sole reason for being is to provide equipment and service to his customers. The complexity of modern machinery causes farmers to rely on responsible dealers for advice on purchase and maintenance of equipment.

8.   Personnel relations.   A retail business is founded on people. The dealership can be no better than its employees. In addition, labor is the largest single item of expense. Put these all together, and it's easy to see that no managerial job is more important than keeping the staff contented and effective.

9.   Public relations.   The manager's image in a community has great affect on the acceptance of the product he sells. Time spent cultivating this image is time well spent.

10.   Sales management.   Selling is not everything, but nothing happens in a retail business until someone sells something. Yes, the name of the game is sell. Whether actively engaged as sales manager or not, the manager cannot duck the responsibility for the organization's sales effort. Selling know-how is a must.

11.   Advertising.   This is a frequently neglected, but very vital, part of any manager's job. The nature and size of the business usually precludes an advertising manager. The general manager must fill in and plan the advertising campaign for the dealership.

12.   Inventory control.   Approximately 60% of the assets are tied up in inventory. Loose control of so much of the business can spell disaster. Suppliers are often over-zealous where your sales potential is concerned. The high-profit dealer is the one who keeps a current inventory and turns it over about three times a year.

MANAGEMENT AIDS AND PROCEDURES

Management is expected to show a profit. No business can survive long without this incentive. The dealer who fails to make a profit really helps no one. The supplier is put in the position of having a potential source of distressed merchandise that must be disposed of in some manner. Lending institutions find themselves with paper of dubious value. Employees are in the position of having to look for other employment. The manager is out of work with a deflated ego. And, lastly, the farmer is without this source of equipment in the community. He is forced to look elsewhere for service and parts for his previous purchases. Failure to make a profit is a real "bummer" with dire consequences for the entire community.

The Break-Even Point

We break even when the gross margin obtained equals the expenses incurred within a given period of time. This is a most important point to consider when setting goals. The goal must exceed the break-even point in order to show a profit. It should be established by the month. Publicize this goal so that all employees are aware of it. A separate break-even point should be set for all departments, then all these be combined into one overall goal.

Reducing operating costs lowers the break-even point. Normally you can expect that for each dollar of cost reduction your break-even point will go down $6.00 or more. The combination of higher margins, lower operating costs, and increased volume will combine to really lower the break-even point.

Table 9.1 shows break-even calculations for all North American dealers for the years 1976–1980.

TABLE 9.1. BREAK-EVEN CALCULATIONS

	1976	1977	1978	1979	1980
1. Sales (000)	1536	1598	1872	2346	2252
2. Gross Profit (000)	276	288	321	395	400
3. Percent (2÷1)	18.0	18.0	17.1	16.8	17.8
4. Operating expense $(000)	211	230	250	309	336
5. Percent (4÷1)	13.7	14.4	13.4	13.2	14.9
6. Profit before taxes (000)	65	57	71	86	76
7. Break-even point (4÷3) $	1172	1276	1458	1839	1887
8. Margin of safety (1−7) $	363	321	413	507	365
9. Margin of safety (1÷8) %	23.7	20.1	22.1	21.6	16.2

Based on "Cost of Doing Business Study," in thousands of dollars.
Gross profit = Margins plus other income.

How then do you make a profit and contribute to the well-being of a community? The accent should be on sales, but volume is not the entire answer. Refer to Table 9.2.

Margins as a percentage of sales are being maintained regardless of volume. Expenses as a percentage of sales were very similar for all except the largest volume dealers. The large-volume dealer showed a profit on sales since his expenses were proportionally lower. The lower-volume dealer managed to keep other income at the same level as the high-volume dealer. This is rather surprising in view of volume discounts offered by manufacturers.

TABLE 9.2 HOW MARGINS AFFECT NET PROFIT (1980)

Category	All dealers Amount ($)	% Sale	Less Than $1,000,000 Amount ($)	% Sale	$1,000,000 to $2,000,000 Amount ($)	% Sale	$2,000,000 to $3,000,000 Amount ($)	% Sale	Over $3,000,000 Amount ($)	% Sale
Total sales	2,252,306	100	720,778	100	1,463,301	100	2,453,181	100	4,813,219	100
Total margins	346,541	15.39	119,025	16.51	224,998	15.38	369,792	15.07	780,337	15.38
Total expenses	335,566	14.89	124,771	17.31	232,728	15.90	370,131	15.08	672,182	13.96
Net profit on sales	10,975	0.49	6,746	(0.79)	7,730	(0.52)	.339	(0.01)	68,155	1.42
Other income	53,746	2.38	16,576	2.29	37,882	2.58	55,338	2.25	114,200	2.37
Net profit or (loss)	64,721	2.87	10,830	1.50	10,152	2.06	55,049	2.24	182,355	3.79

Using the year 1980 as an example, "Break-Even" calculations were run on the four volume categories. They are shown in Table 9.3. The margin of safety is very low for the low-volume dealer. A drop in sales of 7.9% or $57,103 would erode all of his profit. The high-volume dealer has a cushion of 21.3% or $1,026,277. It is apparent that the low-volume dealer must be a better manager than his high-volume counterpart.

Management can be successful only if you are properly staffed and organized. The following ten tests are taken from the Farm and Power Equipment Retailers Handbook.

TABLE 9.3. BREAK-EVEN CALCULATIONS BASED ON SIZE (1980)

Category	Less than $1,000,000	$1,000,000 to $2,000,000	$2,000,000 to $3,000,000	Over $3,000,000
1. Sales	720,778	1,463,306	2,453,181	4,813,219
2. Gross profit	136,601	262,880	425,180	854,537
3. Percent	18.8	17.96	17.32	17.75
4. Operating expenses	124,771	232,728	370,131	672,182
5. Percent ($4 \div 1$)	17.31	15.90	15.08	13.96
6. Operating profit	10,830	30,152	55,049	182,355
7. Break-even point ($4 \div 3$)	663,675	1,295,812	2,137,015	3,786,940
8. Margin of safety ($1 - 7$) ($)	57,103	157,494	316,166	1,026,279
9. Margin of safety ($8 \div 1$) (%)	7.9	11.4	12.9	21.3

Gross profit = Margins plus other income

Ten Tests of Organization and Staffing

1. *Accountability.* Is each department organized around the achievement of a major objective or goal, for which a single executive is held finally responsible and accountable in conformity with definite standards?
2. *Authority.* Is there a clear line of authority running from the top to the bottom of the organization? Does everyone know exactly to whom he reports, what he is accountable for, and what standards he is required to meet?
3. *Progression.* Do all positions within each department, and the company as a whole, provide a natural ladder of progression and increasing scope so that at all times employees are in training for advancements as vacancies occur?
4. *Relationships.* Have all work responsibilities, relationships, and authorities been clarified? Are lines of command, sources of advice, and channels of communication definite and clear-cut?
5. *Effectiveness.* Does the Organization Pattern create a climate that encourages maximum executive performance, effectiveness, and accountability?
6. *Duties and responsibilities.* Are the objectives of each department and the duties and responsibilities of every position within the organization prescribed in writing?
7. *Qualification requirements.* Have qualifications standards for each position been prescribed in terms of specific knowledge, skills, and personal qualities required? How effectively are they being used for the selection and promotion of qualified employees.
8. *Compensation.* Have compensation standards been prescribed in terms of position classifications, grades, and adequate pay ranges applicable to every position in the organization? How effectively are they being administered?
9. *Utilization.* Is each executive and supervisor held finally responsible for the selection, compensation, development, productivity, and morale of all persons who report to him? For the maximum utilization of the highest abilities, skills, and interests of each employee within his jurisdiction?
10. *Performance appraisals.* Is a systematic program of periodic appraisals of executive performance at all levels carried out effectively? Does top management require and use adequate analysis of the results of such appraisals?

Assuming that you are properly organized and staffed, you should be able to manage profitably. Controlling expenses is one of the major functions of management, and salaries are the largest expense. Since the name of the game is sell, the number of employees should be tied to the volume of sales. During 1980, each employee generated an average of $164,884 of sales per year for all dealers participating in the Cost of Doing Business study. The average dealer maintained an overall margin of 15.39% of sales. This varied from $119,025 for the average dealers with sales to $1,000,000 to $224,998 for dealers with sales between $1,000,000 and $2,000,000 and $740,337 for the average high-volume dealer. If margins can be increased, volume may be allowed to decrease proportionately. All labor and salaries should not exceed 65% of the total margin plus other income not due directly to the availability of capital. From this, it is evident that the nonproductive employee cannot be tolerated. The manager must evaluate the productivity of each employee because one excess employee can vary these figures by 8% in the average dealership, where 12 persons are employed. The smaller the volume of business, the more critical is the volume per person. It is problematical whether or not a dealership can survive long with a volume of less than $10,000 per month per employee. This requires margins that are hard to maintain in this competitive business.

Critical Ratios

The athlete uses a stop watch to tell him how well he is progressing. The doctor might use a thermometer or a stethoscope to examine his patient. The business manager uses the financial statement—both the balance sheet and the operating statement—to tell him how well he is doing. It isn't sufficient to analyze the financial statement. It must be done soon enough to affect changes immediately. This should never be delayed beyond the fifth of the month.

It isn't necessary to be an accountant to analyze a financial statement. A good understanding of eight accounts is sufficient to do a good job. The dealer should know what each account is based on. The balance sheet describes the assets, liabilities, and net worth of the firm. From the balance sheet we get the cash, accounts receivable, inventory, notes payable, and accounts payable. From the operating statement we get margins, total operating expenses, and sales. A manipulation of these eight accounts will produce ratios that can be used to evaluate the health of an organization. Ratios are symptoms that, when evaluated, lead to problems that are profit robbers. Follow the Accounts Primer to see the interaction of the balance sheet and the operating statement. The Manager's Account Primer, Figure 9.1, is a convenient way to show the interaction of the balance sheet and the operating statement.

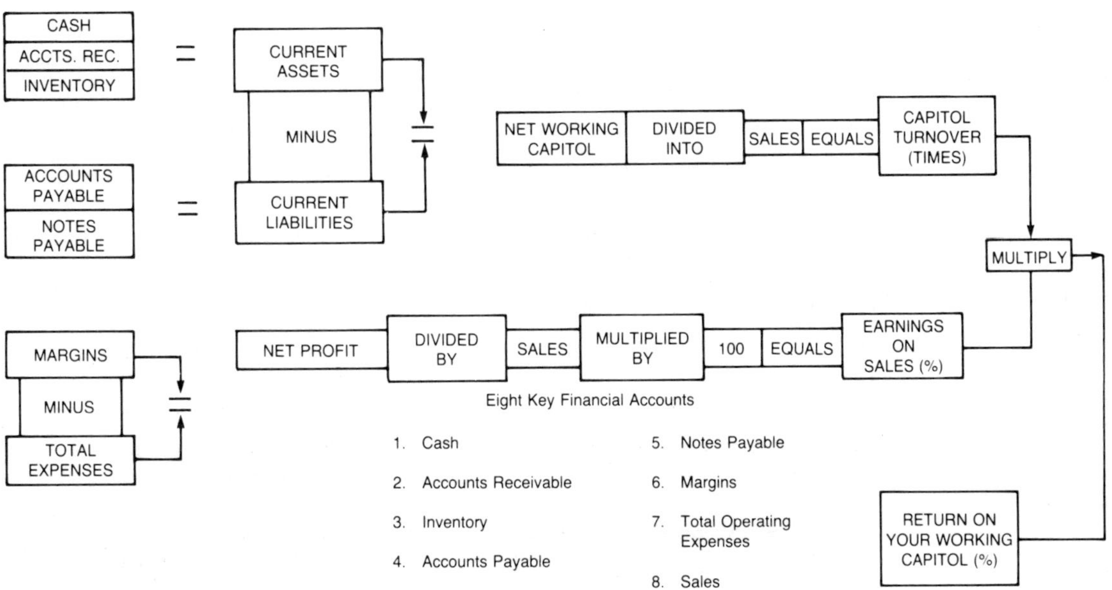

FIG. 9.1. The manager's primer of accounts, from George Scherovich, editor, *Implement and Tractor*

TABLE 9.4   FINANCIAL RATIOS (1980)

Eleven ratios to be guided by:		Current ratio	Ownership equity (%)	Receivables turnover (Times)	Net working capital turnover (Times)
Sales under $1,000,000	Avg.	1.43 to 1	25.68	17.02	5.04
	High	1.60 to 1	36.88	15.42	4.44
	Low	1.34 to 1	16.59	16.59	4.66
Sales of $1,000,000 to $2,000,000	Avg.	1.35 to 1	24.20	18.95	6.98
	High	1.51 to 1	34.15	16.06	5.74
	Low	1.25 to 1	16.18	18.50	8.32
Sales of $2,000,000 to $3,000,000	Avg.	1.40 to 1	27.38	16.98	7.33
	High	1.62 to 1	37.65	15.98	5.52
	Low	1.26 to 1	19.11	17.38	9.39
Sales over $3,000,000	Avg.	1.38 to 1	29.23	15.46	8.66
	High	1.62 to 1	38.77	12.79	6.08
	Low	1.24 to 1	20.12	20.01	11.99

Computed from 1980 "Cost of Doing Business Study." See Appendix A.

Many ratios can be used. Some are always applicable, and others may be used for specialized purposes. They should be used as a guide. They are a goal to shoot at, a position that most dealers find conducive to doing profitable business. Many can be violated repeatedly, but the dealer should understand that he travels at his own risk when he departs from the well-traveled path (Table 9.4).

Ratios may vary considerably according to size of business, location, competition, market conditions, lines carried, and types of organization. Despite these differences, they serve as a guide which prudent managers should watch. Of the 12 listed in Table 9.5, four are most important. Unless success is achieved in these areas, success in the others is not very probable. The four are: (1) capital use efficiency, (2) the liquidity test, (3) return on assets, and (4) inventory turnover.

Remember that the poor ratio is a symptom of something wrong. It is a signal indicating that action is required. The hard part of the manager's job is often to do what the signals direct. In itself a signal means nothing. The manager must locate the problem, diagnose it, and prescribe a cure.

We might describe the management procedure as a three-step process of problem identification:

TABLE 9.5. FINANCIAL RATIOS

Signals	How to figure	What it means
Capital relationships:		
1. Net Working Capital Turnover	Sales ÷ (Current Assets − Current Liabilities)	Capital use efficiency—Capital Turnover
2. Cash and Receivables to Current Liabilities	(Cash + Receivables) ÷ Current Liabilities	Acid test of how liquid you are
3. Current Assets to Current Liabilities	Current Assets ÷ Current Liabilities	Current ratio—your borrowing capacity or pay-out ability
4. Receivables Turnover	Sales ÷ Accounts Receivable	Sound credit and collection policy
5. Ownership Equity	Net worth ÷ Total Assets	% of business owned by proprietor
Profit Relationships:		
6. Return on Assets	Net Profit ÷ (Total Assets	How well total assets are used
7. Net Profit to Sales	Net Profit ÷ Sales	Overall efficiency
8. Gross Margin to Sales	(Sales − Cost of Goods Sold) ÷ Sales	Whether sales are productive
9. Net Profit to Net Worth	Net Profit ÷ Net Worth	Return on your net worth
Merchandise Relations:		
10. Inventory Turnover	Cost of Sales ÷ Average Inventory at Cost	Sales efficiency
11. Inventory to Net Working Capital	Average Inventory at Cost ÷ (Current Assets − Current Liabilities)	Whether too much capital is tied down in inventory
12. Turnover of Total Assets	Sales ÷ Total Assets	Adequacy of capital

Inventory to net working capital ratio	Turnover of total assets (Times)	Return on total assets (%)	Return on net worth (%)	Net profit on sales (%)	Gross margin to sales (%)	Inventory turnover (Times)
2.90 to 1	1.35	2.02	7.87	(0.79)	16.51	1.36
2.19 to 1	1.47	9.87	26.75	3.93	18.73	1.56
3.01 to 1	1.07	(5.51)	(33.22)	(7.02)	14.45	1.04
3.30 to 1	1.62	3.34	13.82	(0.52)	15.38	1.70
2.39 to 1	1.69	11.24	39.92	3.55	17.81	1.88
4.50 to 1	1.45	(3.40)	(21.02)	(4.35)	14.26	1.49
2.93 to 1	1.84	4.13	15.08	(.01)	15.07	2.02
2.10 to 1	1.85	11.22	29.81	3.50	17.40	2.07
4.07 to 1	1.72	(2.38)	(12.49)	(3.40)	13.49	1.89
2.94 to 1	2.03	7.68	26.28	1.42	15.38	2.37
1.97 to 1	1.98	15.42	39.76	4.57	18.36	2.38
4.40 to 1	2.08	0.58	2.88	(1.56)	13.31	2.25

1. Locate general problem area. We do this by studying ratios and records of the total business. We look for symptoms or signals. For the sake of illustration, let us suppose we find overall (1) excessive inventory and (2) low margins.

2. In this step, we try to locate the specific problem. We study the ratios by department. This enables us to isolate the problems. We now determine in this mythical example that we have (1) excessive parts inventory and (2) low service margins.

3. Now, we must diagnose the problem. Not only do we study the records, but we also interview the people concerned if necessary. Why do we have excessive inventory? Is it lack of perpetual inventory system, excessive obsolescence, or just an inefficient and improperly trained parts manager? As important as anything else is what do we do about it? The low service margin may be the result of too low a service rate, improper bookkeeping, or excessive shop labor. The management together with the service manager must identify and rectify the problem.

In summary, the good manager practices management by exception. He compares his records with standards or with his previous accomplishments. He uses these to find the symptoms. He lets these symptoms lead him to the problem and then he diagnoses the problem as it exists. Finally, he does the all-important thing, which is to take corrective action.

QUESTIONS

1. Define management.
2. What are the basic principles of management?
3. Discuss the typical duties that illustrate each of these principles.
4. How does the level of management affect the task of management?
5. Discuss setting goals for an organization.
6. What do we mean by management by exception?
7. Discuss the organization and staffing of business.
8. How can ratios be used to assist management?
9. What are the four most critical ratios?
10. Where would you find the figures necessary to develop ratios for your business?

10

Managing Credit

It's credit, credit cards, or cash. No one will dispute the desirability of cash, but credit and credit cards are a way of life today that many dealers cannot afford to buck. You must realize that credit involves risk, but risk can be minimized when you develop a sound business-like approach to credit management.

Why credit? There are many sound reasons why credit can improve a business. The following are a few of the most obvious:

1. Credit builds a regular clientele.
2. Credit customers buy higher quality.
3. Credit is convenient and builds good will.
4. Goods can be delivered on approval.
5. The credit list becomes a good mailing list.
6. Financial adjustments can be made very easily.
7. Credit creates a more intimate and personal relationship with the customer, which can foster better business relations.

Conversely, we might ask, "Why cash?" This brings out the disadvantages of credit sales that must be taken into consideration.

1. Capital is tied up in receivables.
2. Credit adds to the interest charges the dealership must pay.
3. There is a potential for losses due to bad credit.
4. Some customers may be tempted to buy beyond their ability to pay.
5. Customers tend to return goods and otherwise find fault with machines that are not paid for.
6. Credit adds to the cost of doing business.

When the pros and cons are balanced out, most dealers find that credit of some form is necessary. Management must apply the proper safeguards to make it profitable.

TYPES OF CREDIT

All credit cannot be lumped together. Although anything other than cash is normally called credit, the type of credit plays a vital role. The credit card, the installment credit, and the open charge account all offer different degrees of security and return to the dealer.

In installment credit the merchandise is normally secured with some collateral. This may well be the equipment itself. Interest is charged based on prevailing rates. This type of paper normally can be used as dealer collateral for short-term loans or sold direct to the bank. There is one cardinal rule to follow when extending installment credit. "Obtain sufficient down payment to protect the seller and to give the buyer a sense of ownership. Insist on monthly or frequent payment to more than compensate for the use of the machine." The installment contract has several advantages over the open charge account:

1. Security, as the dealer has a lien.
2. Interest is charged.
3. The paper is negotiable.
4. The customer will pay the interest-bearing note sooner than the open account.

The credit card is becoming widely acceptable to most people. This appears to be an excellent means of handling receivables. For a fee varying from 3 to 5%, a number of companies will handle your credit. This is on a nonrecourse basis: The credit card company agrees to absorb all losses due to nonpayment.

Many local banks will also "bankroll" your receivables. It is normally necessary to get all the dealers in an area to band together for this purpose in order to make it profitable for the bank to set up for this service. This service may be on a nonrecourse, recourse, or partial recourse basis. It is very important to understand the terms of the contract.

A study made by the John Deere Company in 1972 indicates that the cost of handling receivables varies from $1.54 to over $16.00 per hundred dollars depending on age. There has been a 50% rise in the interest rates since then and probably as much in other costs involved. The weighted average is estimated to be about $6.00 per hundred dollars.

Changing over to a credit card system may pose several problems if an individual dealer does it. Experience indicates that the transition can be smoother if it is made by several dealers in an area at one time.

Most national cards do not handle cards for corporations. These still have to be handled individually by the dealer. The dealer should assist customers to obtain cards by helping them fill them out. It is very important to set limits on the card that will allow the customer the amount of credit he needs. This prevents having to check with the credit card company when limits need to be exceeded.

Credit cards also pose problems. The question of lost cards always comes up. Under the "Truth in Lending Law" the customer is protected against all losses over $50.00. The dealer is not held responsible for bad cards when he has no way of knowing the cards are bad. When a dealer is notified of a bad card, he should mark the date of receipt of the warning letter.

When picking up machines off the farm for service, it is wise to ask for the credit card at that time. On phone orders, the number can be written in and the order signed by the parts man or service manager. In other cases where the card or signature is not readily available, the customer should be billed by letter immediately.

Servicing open credit accounts poses many problems. With proper management, these problems are not insurmountable. Start out with a credit policy. A credit policy is a business management tool to aid the manager in controlling the extension and collection of retail credit. Credit itself is a sales tool, but a credit policy is a management tool.

The credit policy should have an objective. Normally this may well be to minimize the cost of extending credit and therefore to maximize profit. Some dealers may have other objectives as well.

A credit policy should have at least three basic elements. It must outline a procedure for:

1. Extending credit
2. Controlling credit
3. Collecting credit

The credit policy should be a written document. All three elements of the policy should be detailed. Post it conspicuously in your place of business. Let all your employees know what it is. By all means, let your customers know what your credit policy is. Feelings are hurt and customers lost when your credit policy is sprung on them after a transaction is made.

EXTENDING AND CONTROLLING CREDIT

Appoint a credit manager for your store. Give him the authority to extend or withdraw credit privileges. Let him award credit on the basis of an application for credit. These applications must be screened and references checked when the customer is not well known.

Set limits for the account. When new accounts are opened is the best time to agree on how much credit will be needed, for how long, when it will be repaid, and what income will be used to repay. This is the best time to agree on carrying charges. Remember that the "truth in lending law" requires you to divulge any charges made on the account. Make notations on the application as to how the customer meets his credit obligations.

Your policy on controlling credit must start with the handling of sales tickets. Have them numbered sequentially. Never destroy but rather void any unused tickets. See that all tickets are accurately posted to accounts receivable.

TABLE 10.1. AGED ACCOUNTS RECEIVABLE REPORT[a] Page No. 1, Date , 19

Name	Balance (credits in red) ($)	Current[b] (Superior) ($)	30 days past due (Excellent) ($)	60 days past due (Good) ($)	90 days past due (Fair) ($)	Over 90 days past due (Poor) ($)	Action
Alford, John H.	35.00	35.00					
Akerly, John I.	60.00	20.00	40.00				
Anderson, Bill D.	165.00	100.00	25.00	40.00			
Apples, Jack K.	110.00	20.00	15.00	30.00	45.00		
Azar, Tom K.	150.00					150.00	
Bracken, Jim C.	125.00	20.00	35.00	15.00	25.00	30.00	
. . .							
Landry, Henry L.	0						
Martin, Sonny L.	75.00	75.00					
Zubar, Donald L.	100.00			100.00			
Total	8,075.00	2,915.00	2,510.00	950.00	800.00	900.00	

[a]Terms: Net 10 days from date of billing (20th of month).
[b]Charges made in the period from the previous date of billing to the current date of billing.

Have your bookkeeper age your accounts receivable and have this report in the hands of the credit manager three days before mailing the monthly bills. An aged accounts receivable report may take many forms, but as a minimum it should approximate the one shown in Table 10.1. If your accounting is computerized, this form should be prepared for you by the computer from the daily entries keyed into the memory. If an action column is not provided, draw one in.

The first step in collection is for the credit manager to fill the action column. A current account can be classified as superior and needs only to have a monthly statement sent out (Action 1). The account that is 30 days past due may be considered excellent. A reminder statement should be sent out with the monthly invoice (Action 2). It should be short and courteous. It may take the form of a stamped statement on the bill indicating the account is 30 days overdue.

An account that is 60 days past due is certainly good but needs attention. A reminder letter (Action 3) is in order. Again it should be short, courteous, and to the point. The customer should be made aware that you know his account is overdue. This will give him an opportunity to pay up or come in and explain his delinquency.

An account that is 90 days past due might be classified as fair. This calls for an appeal letter (Action 4). As the name implies, you appeal to his sense of fairness and understanding. Depending on the circumstances, it might be well to make a personal phone call to the individual. This should give you an opportunity to size up the situation. Why hasn't the customer paid? Does he disclaim the bill, think he was overcharged, or that the merchandise is unsatisfactory, or is he just strapped at the moment? It is well to anticipate any of these circumstances ahead of time so that a meaningful dialogue may be carried on. The personal call might be indicated as PC.

Accounts that are over 90 days past due require that you get tougher. Remember, there is nothing wrong with asking for your money—it is your money you are asking for. A personal visit (PV) is certainly needed here. Armed with all the information and circumstances surrounding the case, sit down with the customer. Reach an agreement. With a failure to reach an agreement you should follow with a "threat to legal action" letter (Action 5). This in turn is followed by the attorney's letter (Action 6). It is important to follow through when these threats are made.

When credit is properly extended, very few if any cases will go to stage 6. Somewhere along the way they are either paid or settled by special arrangement (SA). When an account has been granted a special arrangement due to circumstances beyond the customer's control, care should be taken not to continue to badger the customer. You will monitor to see that he lives up to his agreement.

It is often difficult personally to call or visit some of your customers. The need to visit can often be eliminated. Additional credit is normally the result of parts purchase or the purchase of service labor. Furnish the heads of each of those departments with a copy of the aged accounts receivable document. A dotted line drawn through a customer's name might mean that he is to be given further credit but asked to visit with the credit manager on the way out. A solid line drawn through the customer's name might mean he is to receive no further credit until the credit manager sees the customer and reaches an agreement. This procedure can be effective but records must be kept current at all times.

The same aged accounts receivable document is repeated in Table 10.2 with the action column filled out. The credit manager pencils in the actions, and the bookkeeper inks them in as they are completed. Visits and calls should be inked in by the manager when they are completed.

Name	Balance (credits in red) ($)	Current[b] (Superior) ($)	30 days past due (Excellent) ($)	60 days past due (Good) ($)	90 days past due (Fair) ($)	Over 90 days past due (Poor) ($)	Action [see text]
Alford, John H.	35.00	35.00					1
Akerly, John I.	60.00	20.00	40.00				1, 2
Anderson, Bill D.	165.00	100.00	25.00	40.00			2 2,3
Apples, Jack K.	110.00	20.00	15.00	30.00	45.00		2, 3 2, 4, PC
Azar, Tom K.	150.00	CREDIT O.K. TO $100 10/15/64 J.P.D.				150.00	2,4, PC 5, PC
Bracken, Jim C.	125.00	20.00	35.00	15.00	25.00	30.00	SA, 1 SA, 1
. . .							
Landry, Henry L.	0						1
Martin, Sonny L.	75.00	75.00					1 1
Zubar, Donald L.	100.00			100.00			2 2, 3
Total	8,075.00	2,915.00	2,510.00	950.00	800.00	900.00	

[a]Terms: Net 10 days from date of billing (20th of month).
[b]Charges made in the period from the previous date of billing to the current date of billing.

In summary we see that we start controlling credit with the credit application. The accounts receivable ledger and the aged accounts receivable report are the primary documents we rely on. The key people who must take responsibility and be given authority are the credit manager, the bookkeeper, the parts man, and the service manager. Together they form a team that must bear the brunt of the load.

The Farm and Power Equipment Dealers Association through their monthly magazine lists the relationship between age of accounts receivable and the chance of collecting them. From Figure 10.1, it is evident that an active collection policy at the beginning pays good dividends.

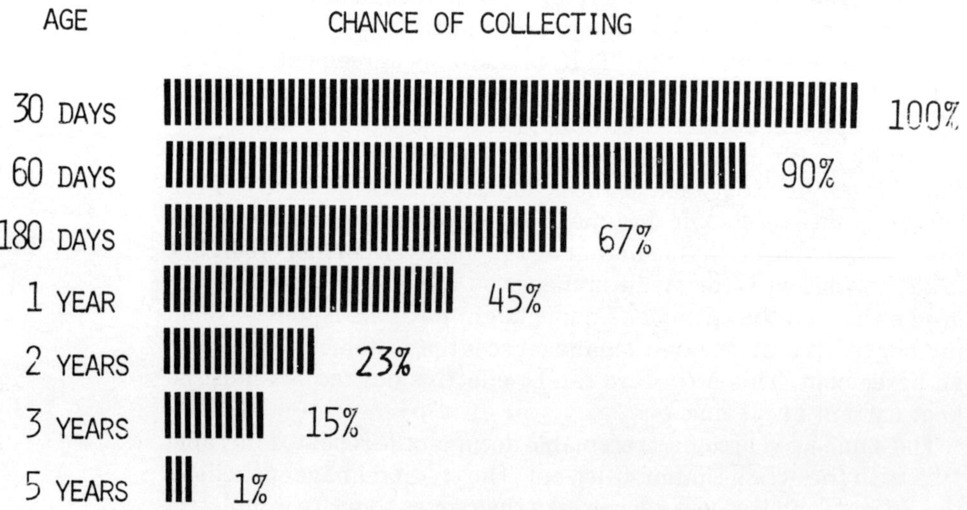

FIG. 10.1 The relationship between age of accounts and your chance of collecting.

EVALUATING YOUR CREDIT POLICY

Your credit policy can be evaluated using the budgetary approach. Analyze and determine for yourself what credit costs are. They should be reduced by whatever interest charge are realized. Determining the increased sales due directly to credit sales is much more difficult. Credit has become such an accepted thing that perhaps the most practical solution is to consider it a cost of doing business and keep it at acceptable levels. There are several industry ratios that may be helpful in determining what is an acceptable level.

Some Standards

1. Receivable turnover. This is computed by dividing total sales by receivables. A low ratio might indicate collections are not up to par. Since they may vary by season, it is well to compare monthly figures with previous months. During 1980 the Cost of Doing Business study showed that receivables turnover varied from 13.87 to 18.51, with all dealers averaging 16.59 times.

2. Monthly sales parts and service to receivables. Accounts receivable should not exceed two months' sale of parts and service.

3. Bad debt to total open account sales. This will show the effectiveness of your operating controls and collection procedure. Bad debts are hard to define. Perhaps any debt over 180 days should be placed in that category for the purpose of computing this ratio. Experience may dictate some different figure.

4. Collection period. Determine the average daily credit sales, which is the annual credit sales divided by the number of days per year. Divide the current accounts receivable balance by the average daily credit sales. This will give you the number of days of credit sales. If you exceed your goal by more than 10 to 15 days you have cause for alarm. A goal of 60 days is a good rule of thumb to follow.

Let us apply these standards to "All Dealers" as shown in the 1980 Cost of Doing Business study:

1. $\text{Receivables turnover} = \dfrac{\text{Total sales}}{\text{Receivables}}$

$$= \dfrac{2,252,306}{135,745} = 16.59 \text{ times}$$

2. Monthly sales parts and service to receivables:

$$\dfrac{\text{Receivables}}{\text{Monthly sales parts and service}} = \dfrac{135,745}{(395,027 + 149,875)/12}$$

$$= \dfrac{135,745}{45,408} = 2.99 \text{ months}$$

3. Bad debts to total open account sales:

$$\dfrac{\text{Bad debts}}{\text{Receivables} \times \dfrac{\text{collection period}}{\text{days in year}}} = \dfrac{\$3,313}{\$135,745 \times \dfrac{89}{365}} = 0.59\%$$

4. Collection Period. The Cost of Doing Business study does not show enough data to determine this ratio. You may find yours by dividing your accounts receivables by your average daily credit sales. This gives information similar to the second ratio shown in this list.

An analysis of these ratios show that the average dealer loses 0.59% of his receivables annually to bad debts and carries his open accounts customers for 89 days before collecting. This is no improvement over 1979, when bad debts were 0.61% of open account sales and the collection period was 89 days.

Receivables are a sales tool and can be a profit generator when properly managed. During 1980, the average high-profit dealer had a receivables turnover of 18.51 times. This is contrary to what is normally expected. The relaxed credit policy evidently stimulated sales and profits.

Notes

Notes receivable are promissory notes accepted by you on merchandise sold. They may also be used to secure past due accounts receivable or to cover extended terms not normally offered. The promissory note is a legal instrument by which one party (a customer) promises to pay the second party (the dealer) a certain sum of money at a stated future date. Notes on past due accounts should be secured by some personal property if at all possible. Security could be machinery, crops, stocks or bonds, and the like.

Since notes are secured and bear interest, the customer is encouraged to pay promptly. The dealer has the option of foreclosing if the terms are not met. This is always unpleasant and should be avoided when possible. The note has the added advantage that it can be sold (at some discount rate) to the bank or other lending institution. Discounts for this service vary. The need for operating cash must be considered. Notes receivable should be listed on the balance sheet and managed as any other receivable.

Summary

Credit selling can increase sales and profits when properly administered. Good management of accounts can be accomplished only if good records are maintained. Base your credit decisions on knowledge of the customer. Set up a collection system and follow it regularly. Always keep in mind when collecting that you want to:

1. Get your money.
2. Keep your customer happy.
3. Train your customers to pay promptly.

QUESTIONS

1. Discuss the advantages and disadvantages of credit sales.
2. What costs are involved in open credit sales?
3. What are the three elements of a credit policy?
4. What is the "Truth in Lending Law"?
5. What basic document is used in extending credit? In controlling credit?
6. Discuss the action column of the aged accounts receivable report.
7. How can you evaluate your credit policy?
8. What should you keep in mind when making collection efforts?

11

Parts Management

REPAIR PARTS IN RETAIL BUSINESS

Repair parts account for an average of 17.54% of the total sales volume of all North American dealers according to the 1980 Cost of Doing Business Study. That same document reports that the average margin was 30.8% of sales. Occupying about one-sixth of the total floor space, this department accounts for 35.1% of the total dollar margin. It is easy to see why parts management is so important.

The figures quoted represent average values. Margins are not always as high as this. Let's look at some of the conditions that lead to low margins.

First, we have lack of control. Parts are literally given away, with no regard of the transaction. Customers realize that, by complaining loudly, they can obtain free parts. It is often necessary to make adjustments. When records are kept, these adjustments can be recovered from the supplier.

Second is repairing used equipment. Internal repairs of used equipment for the dealership often result in parts not being properly charged out. In some cases this may only result in bookkeeping errors. Often, however, the true cost of used equipment is not known, nor it is ever recovered. Good management demands that books reflect true conditions. Unreported parts sales result in lost profits and erroneous inventory figures.

Third is failure to recognize obsolete parts. Poor records, changing sales patterns, and unclaimed orders often result in parts that no longer move. Suppliers redeem these errors in part if the dealer recognizes their obsolescence in time and returns them for credit. This can only happen if a policy on obsolescence is in effect.

Fourth is discounting parts sales. Be it to neighboring dealers, large customers, or personal friends, the result is the same. The realized margin will be less than anticipated.

Fifth is the failure to add freight and handling charges. Sales price has to recover invoice cost, freight, assembly and handling, taxes, and overhead before a net profit can be realized. Special orders requiring long distance calls and special handling always reduce drastically the potential gross margin. When figures are not available, add 10% for freight.

The objectives of any good parts department are usually four-fold. They may be listed as follows:

1. To have parts available when needed by customers
2. To keep parts safe and secure
3. To assure that parts sales are profitable
4. To keep obsolescence to a minimum

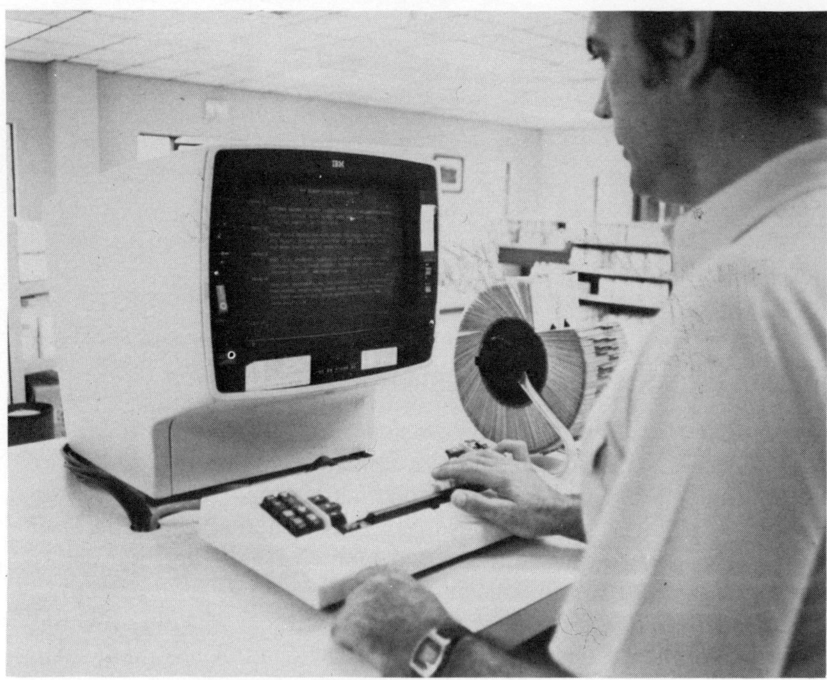

FIG. 11.1. Modern parts departments let the computer write the sales tickets and enter the transaction in the central records.

The parts manager's job is to accomplish these objectives. It starts with inventory control—knowing what is on hand, what it costs, and how it moves (Fig. 11.1). Knowing how it moves is the basis for the stocking plan. Good merchandising will do the rest.

INVENTORY CONTROL

For years the heart of inventory control was the stock record card. The computer has done much to change the shape of the card. Its message, however, remains much the same. Regardless of whether it is recorded manually or electronically, the information needed is basically as follows:

1. Identification of the machine the part is used on
2. Substitute or replacement part number
3. List and net price
4. Part's sales history
5. Date order place
6. Quantity ordered
7. Whether order received
8. Sales against inventory
9. On-hand inventory after sale
10. Bin location
11. Stocking level and reorder point
12. Inventory at a given date
13. Annual inventory on hand
14. Return sales—inventory adjustment

It's easy to see why computerized inventory accounting is so popular. It replaces manpower, reduces cost, and increases profits. With the card system, sales slips have to be posted to the cards daily in order to have a perpetual inventory. It's a tedious job that is very easy to get behind on. Colored flags or other codes are used to indicate stock that is in need of reorder.

The computerized system does not entirely eliminate the parts man's tedium, but it does reduce it. A pad is kept on the countertop, and sales are posted to it as they occur. At stated intervals they are mailed to the computer center for entry into the master file and for bringing up to date the parts history and computing reorder quantities. Generally bimonthly, the dealership's order of parts is printed out by the computer and sent to the dealer for acceptance and subsequent forwarding to the supply depot. It can be altered by the dealer prior to mailing out. It is most important that daily lost sales be posted to the pad. This is the only way that the computer can add items to the stocking list. Dropping parts from the stocking list poses no problem, as the computer can be programmed to sense lack of sales of a stocked item.

Supply procedures and policy vary between part suppliers. Not only do they vary between suppliers, but they also vary with time. Procedures vary to facilitate restocking, to encourage certain buying habits, to discourage others, and generally to increase parts supply efficiency. Basically, however, the procedure between companies is very similar.

Suppliers encourage a basic order of parts on a twice-a-month basis. They encourage this by paying the freight on these basic orders. They will honor orders at any time, but at an increased cost to cover the special handling. Dealers are assigned a reorder date to balance the workload of the supplier.

Four seasonal orders are usually solicited on a favored basis. These cover orders for parts for specialized equipment used only for a short period of time, such as cotton pickers or combines. Dealers try to stock 80% of their past year's sales, then depend on the bimonthly orders to fill in any shortages that may develop.

The parts man constantly is placed between the customer who wants his part always immediately on hand and the management that insists on running a profitable parts department. Management must make the decision as to what percentage of orders can be filled on call. Ninety percent is considered good parts service. Once the decision is made, the parts man must keep the inventory in balance. A level of supply is established. It may vary from a not less than 45-day supply but not more than 90-day supply to a 30−60 day basis. The basis on a local level has to be based on reorder time from the supplier. The turnover from a 30−60 day basis will be higher than the 45−90 day basis. The percentage of orders filled will be lower.

An example of how to set reorder points and to reorder quantities may serve to illustrate the mathematics involved. Assume a 15-day lead time from the warehouse and an item with 24 sales per year. Ninety days' supply is six items. This item should be reordered when the stock level reaches four, or a 60-day supply. At this time three items should be ordered. These should be received in 15 days, at which time the level should be down to 45 days' supply or three. When the three from the current order are received, the total on hand becomes six, or a 90-day supply, again. Sales are seldom that uniform. However, establishing reorder points and reorder quantities simplifies the reorder procedure and, if demand is monitored carefully, goes far toward warding off parts obsolescence.

If a 30–60 day stocking level is practiced, the procedure for an item with 24 sales per year would be as follows: The maximum on-hand quantity would be 60 days' supply or one-sixth of 24 or four items. The reorder point would be three. At that time two items would be ordered. When they are received in 15 days, the stock level should be down to two. This would bring the level back to four, or 60 days' supply. This system will maximize turnover but increase the need for special orders and phone calls. This results in increased cost and lower margins. A balance must be struck where profit is maximized. After all, profit and not turnover is the final criterion.

Generally, it is not profitable to stock items that sell less than two per year. These should be marked nonstock and ordered on demand. Low-priced items such as nuts and bolts should be ordered in quantity. A four-month supply or $20 is a logical order.

The parts man has to be a trend watcher. Stock levels must be constantly updated. These stock levels should be set on the last 12 months' sales. When the trends over several years is down, the level should be reduced below the last year's sales level to compensate. Seasonal parts are a particular hazard. Large carryover has to be minimized.

Stocking parts for new machines is always a gamble. On items selling for less than $50, stock when you get at least two calls. Compare with similar machines. Manufacturers' representatives usually have suggestions as to what to stock.

Whether based on a stock record card or on a computer printout, the stock order has to be mailed in. Each supplier has dates and times and methods to order to get the lowest rates. Follow them. As a minimum, the order will show quantity wanted, parts number, and bin location. As important as the initial order is what to do when it is received.

Check the loading tally and the freight bill. Do the weights agree? Did you get all the boxes or cartons? Are any cartons broken? Check the correctness of the name, address, and shop. Check the freight rates. Upon opening the cartons, check the shipment against the packing slip and against the order. Bin the parts, then post on the stock record cards if you are on the manual system or on the pad if you are using the NICS system. Advise the supplier if parts were shipped in error. Normally they should not be returned until shipping instructions are received. Keeping parts shipped in error is often a cause of parts obsolescence and low turnover.

Since turnover is an indicator of good parts management, it is well to consider its significance as well as how to compute it. Turnover for any given period is the cost of sales divided by the average inventory at cost for a given period.

$$\text{Turnover} = \frac{\text{Cost of sales}}{\text{Average inventory at cost}}$$

During the year 1980 the turnover of the average North American dealer was 1.84. It varied from 1.52 for the under $1,000,000 dealer to 1.77 for the dealer with volumes between one and two million dollars per year to 2.08 for the dealer with over three million dollars of annual volume.

Table 11.1 makes an interesting comparison between dealers in various volume categories. This table was taken from the 1980 Cost of Doing Business Study (Appendix A).

The average dealer failed to reach the desirable two-to-one turnover. The low-volume dealer did not reach his objective. It is evident that turnover increases with volume and does not reduce the percentage of total inventory in parts. By current standards, the dealer in the under $1,000,000 sales category is hard pressed to meet competition.

TABLE 11.1. THE PARTS DEPARTMENT: A STUDY OF ASSETS AND PARTS TURNOVER, 1980

Type of dealer	Sales under $1,000,000		Sales $1,000,000– $2,000,000		Sales $2,000,000– $3,000,000		Sales over $3,000,000	
	% of Inventory	Turnover	% of Inventory	Turnover	% of Inventory	Turnover	% of Inventory	Turnover
All dealers	17.4	1.52	15.9	1.70	17.5	1.67	15.9	2.08
High-profit dealers	15.4	1.84	15.9	1.88	16.9	1.91	17.2	2.28
Low-profit dealers	17.2	1.27	15.3	1.49	17.1	1.63	14.3	2.04

PARTS DEPARTMENT LAYOUT

The parts department should be laid out with the service counter in plain sight. Usually parts are openly displayed but not accessible to the customer. There is available a good selection of bins for proper storing of parts. Bolted or quick-connected shelving can be used. A standard height of 7′3″ is widely used. They are available in heights of 6, 8, 9, and 10 feet. Shelving varies in depth from 12 in. to 36 in. It may be spaced as needed, usually in one-inch increments. Shelf dividers, partitions, label holders, bin fronts, shelving boxes, doors, and inserts are all available to complete parts storage.

Plan your layout carefully. Store fast-moving parts close to the counter. Keep the aisles free. Provide good lighting. A receiving area is a necessity. Remember to provide for the "heavy" parts and the odd-shaped parts. Plan a simple shelf coding. Do not use numerical sequence based on parts numbers. This works only for a short while.

FIG. 11.2. A place for every part and every part in its place.

The parts department needs quick access to the bookkeeping office. The service shop needs ready access to parts. Have a window available to serve the shop. It is seldom desirable to have mechanics picking parts.

Parts catalogs should be easily accessible on the counter. Stock record cards must be kept where they are available for quick reference and for posting. Machine manuals are needed for parts reference. They assist in selling kindred items.

MERCHANDISING PARTS

It is most important to set a goal for the department. The goal must be attainable. It should be measurable. There may be more than one goal.

Dollar volume is a popular goal because it is easily measured. Based on past years' experience, reasonable growth rates can be set. It is very important to have concurrence from the parts manager in setting the dollar volume. It then becomes his goal, not just a company goal.

Volume itself is not the whole answer. Sales must be profitable. A dollar volume goal without a percentage margin goal can be meaningless. To be inclusive, turnover should be a goal as well. A good turnover rate means good return on investment. In order to increase turnover, it is essential that stock levels be minimized. A goal for obsolescence may be helpful. To keep customer satisfaction in balance, the goals should include a figure for percentage of parts filled from stock.

In summary, we see that several goals for the parts department may be established.

1. Annual sales volume
2. Percentage margin on sales
3. Percentage obsolescence
4. Turnover of parts

FIG. 11.3. Attractive displays promote impulse buying.

What is obsolescence? It is difficult to define in all cases. A policy should be established for a dealership. That policy must include a systematic method of comparing manufacturers' "eligible for return list" to the dealer's list of parts that have not moved in the past year. A common practice in the industry is to allow a return of selected parts equal to 5% of the year's sales. The percent refund by manufacturers varies but is usually in the 90% or more category. If the part is not listed and has not moved in two years, it should be scrapped or otherwise disposed of. To keep it on inventory inflates your financial statement. Recent rulings of the courts no longer allow you to maintain scrapped parts in your bins for possible future sales. Material declared obsolete must be removed from bins. Net parts obsolescence should not exceed 0.5% of parts sales. If it runs over 1% of parts sales, you have room for real concern. By carefully monitoring individual parts, sales, and manufacturers' "eligible for return list," parts turnover can be maintained at 2.5% and obsolescence maintained under 0.5% of sales.

The parts man must be more than an order taker and stock record clerk. He must be a salesman. Every order is an opportunity to sell related items. This, in the final analysis, is a service to the customer. It saves him a trip back because he forgot to get the gasket, lock washer, or associated part that will break in the near future. The parts man must be trained in this technique and school himself to know the related items.

Advertising plays an important role in parts sales. Push those items that can be acquired through many sources such as sweeps, sickle sections, plow shares, filter elements, spark plugs, hoses, and batteries. Very little can be gained by advertising slow-moving parts and others such as structural parts that have no other local source of supply.

Advertising should consist of point-of-purchase advertising, outside advertising, and special promotions.

Point-of-purchase consists of displays within your place of business. Promote impulse buying. Display parts in an open attractive manner (Fig. 11.3). Group them in relation to related items. These displays will remind your customer of the things he needs. Tie these needs to various parts and service sales programs. Time these displays well in advance for seasonal items. Do not let your displays become stagnant and accumulate dust. Change them often.

The trend is toward self-service merchandising on a supermarket scale. Dealers report good results from this method of merchandising. It does require special display equipment. International Harvester Company in their dealer management literature has this to say about where and how to display parts:

Your point-of-purchase displays normally consist of tri-plane and peg board tables, special display racks, the parts counter, bin ends, display floor windows, and floor displays. These must be located where they will be seen by the greatest number of customers. The basic reason for displays is to put merchandise where your customers can see it and serve themselves. Experience has shown that certain procedures produce better results than others. Let's discuss some of these.

Special racks, tri-plane and peg board tables are excellent for displaying impulse items such as hydraulic components, grease guns, accessories, fan belts, filters, seat pads, mufflers, and other competitive items. Much of your fast turnover merchandise is adaptable to display selling. Customers like to handle the merchandise they buy; they find it interesting. These displays are easy to set up and maintain. They enable your partsman to handle a larger volume of sales without additional help.

Bin end displays are more successful when used for familiar parts, such as plowshares, spark plugs, and other seasonal items. It is a good idea to put reminder items where they command the customer's attention. The items that you want to push should be placed on shelves at eye level.

Your parts counter is a good place to display new items, or advertised daily or weekly specials. It produces the best results with impulse buyers. One word of caution—guard against too much cluttering on counter displays.

Most dealers find floor displays are best for large items such as barrels of hydraulic oil, twine, and deluxe seats or for large quantities of small items in a special promotion. A floor display should be well stocked, safe, and at a convenient height.

Sales floor window displays have one definite purpose: to stop the customer and bring him into your place of business. They must be appealing and attract attention by action. This necessitates the use of special signs, background lighting and display cards. Window displays are most effective for whole goods, or for tie-ins with advertised promotions.

Some Proven Principles of Effective Displays

In point-of-purchase display merchandising you should:
1. Display in quantity—a single item does not attract customers. You will find bulk attracts attention and gives the impression of demand.
2. Use proper arrangement—consider the traffic pattern. Locate your displays where the greatest number of people will pass them. Each table, shelf, etc., should have a relation to the others. Establish a "theme" for the entire area, or by table, such as "Tune-Up for Spring," or "Winterize Now," etc. Do not display plow bottoms, anti-freeze and cultivator sweeps on the same shelf or table. This only confuses the customer. Keep all displays well stocked; remove a few items from the cartons, and do not make the display so orderly that it will discourage the customer from picking up items.
3. Change your display frequently—as your customer's parts needs change, so should your displays. Many dealers establish a rule to change displays after a given number of weeks. This practice permits more parts and accessories to be displayed and helps keep them current with the season. Even displays of non-seasonal parts should be changed because customers do not like to see the same thing on each visit. The change takes very little time, and you will find that it improves the appearance of your department and pays off in additional sales.
4. Keep displays clean and neat—this is a problem at many dealerships, but all agree good housekeeping is necessary. Dealers who have clean and neat displays make someone responsible for keeping them that way. If you give your partsman the responsibility of cleaning the area once a day and follow through to see that it is done, you will have clean, neat displays.

Point-of-sale display advertising takes advantage of the customer's presence at your store. Plan your displays so that they will act as silent salesmen.

Outside advertising will take the bulk of your advertising budget. It may consist of newspaper, direct mail, parts catalogs, radio, TV, road signs, and handbills. Details of this type of advertising are common to all sales and will be covered in the chapter on advertising.

Special promotions complete parts advertising. This method is limited only by your imagination. Parts sales can be promoted with "overhaul" specials. List the parts needed for various jobs and their dollar value. Feature special days. This is an opportunity for store-wide sales promotion. Whatever you choose to do, remember that you must attract buyers.

SUMMARY

A properly managed parts department can be highly profitable. Let's look at some of the indications of good management.

Sales per parts man must be considered. Labor is a major portion of parts department overhead. According to the 1980 Cost of Doing Business Study, sales per parts man varied from $138,510 for the dealer in the under $1,000,000 sales category to $172,037 for the dealer in the $1,000,000 to $2,000,000 sales category to $187,478 for the dealer in the $2,000,000 to $3,000,000 annual sales and $211,895 for the over $3,000,000 total annual sales. This pace can be maintained only in a parts department that is well organized physically.

Dealers should expect parts volume to average about 18% of total sales. When this drops below 15% there is reason for concern. The drop could be due to a sudden increase in whole goods sales. Parts sales will catch up later. It could also be due to someone else selling your parts. When parts sales become greater than 18% of sales, it indicates that whole goods sales may need additional effort. There is cause for concern when parts sales exceed 20%. Since only 30–35% of parts are sold through the service shop, it is to the dealer's advantage to sell whole goods rather than extra parts.

Margins are a vital consideration. They should be maintained at not less than 25% of sales. It is seldom possible to exceed 30%. Turnover results in profit on investment. Low-volume dealers suffer in this category. The rate of turnover for dealers by sales volume category is shown in Table 11.2.

TABLE 11.2. HOW VOLUME AFFECTS PARTS DEPARTMENT PROFITABILITY

Category	Total Sales $	Parts Sales $	Percent Total Sales	Parts Turnover	Parts Margin %	Percent Total Margin
All	2,252,306	395,027	17.5	1.84	30.80	35.1
Under $1,000,000	720,778	160,671	22.3	1.52	31.57	42.6
$1,000,000 to $2,000,000	1,463,306	278,700	19.0	1.77	30.18	37.4
$2,000,000 to $3,000,000	2,453,181	419,949	17.12	1.69	30.69	34.8
Over $3,000,000	4,813,219	786,132	16.33	2.08	31.03	32.9

There is a steady increase in turnover with volume. The high-profit dealer maintains a higher rate of parts turnover. The margin as a percentage of sales is maintained regardless of volume. It contributes to the financial success of all dealers regardless of volume, but it is particularly important to the small-volume dealer. No dealer can stay in business long unless he operates a well-managed parts department.

QUESTIONS

1. When margins are lower than anticipated, what should we look for?
2. What should be the objectives of a good parts department?
3. What is the basic document used to control parts?
4. What is a reorder point?
5. How do you compute parts turnover?
6. What is a "point-of-purchase" display?
7. In analyzing a parts department, what key ratios would you look for?

12

Managing the Service Department

The service department sells labor. There is really little difference between selling labor and selling other commodities. The only basic difference is that labor has zero shelf life. It must be sold hour by hour as it is bought. Unless this fact is well understood, it is not possible to run a profitable shop operation. Every hour unsold is wasted forever. Shop management, therefore, must be constantly on its toes.

THE SERVICE MANAGER

Good management demands the ability to organize resources and personnel and to direct their action to accomplish a certain mission. An efficient, profitable service shop must have a manager. The manager's ability and dedication will go far toward making the shop profitable. He must certainly be mechanically qualified, but more important is his leadership and managerial ability. The commodity he sells—labor—requires attention to detail and both a mechanical and a personnel background. He must be able to delegate responsibility. His responsibilities are many. The following are the primary ones:

Satisfy customers (often the most difficult task)
Direct employees
Control expenses
Select and train shop employees
Hold service meetings
Maintain or replace service equipment
Select and install new service equipment
Process time cards, work orders, and service records
Promote labor sales
Improve department efficiency
Supervise safety activities
Direct set up and delivery of equipment
Maintain profitability of shop

The service manager is the key to the service shop. We've talked about his responsibilities. Now let's look at what desirable characteristics he should have:

1. He must have good character and enjoy the full trust and confidence of the owner or manager.
2. He should know the retail equipment business and have strong loyalty to the dealership and to the line.
3. He must have mechanical experience and aptitude.

4. He must have a sympathetic understanding of people when their equipment is not working.
5. He must be able to listen to customers, to employees, and to management.
6. He must command the confidence of customers in his diagnosis of their equipment ills.
7. He must understand and enjoy selling service.
8. He must appreciate the need for and have the ability to keep records.
9. He must be able to deal with all types of people.
10. He must have the ability to train and instruct.
11. He must command the respect and confidence of the men he supervises.

PERSONNEL RESPONSIBILITIES

Service management is responsible for the selection, hiring, and training of the employees in the service department. The selection of employees must be done with care. The wrong man can cause the loss of customers, loss of retail sales, loss of service sales, and even loss of salary paid to the service man. Investigate before you hire. Don't be like the manager who called the new employee in and said, "Tell me again how good you are. I'm beginning to have my doubts." Employ and train a well-balanced team that will draw trade to the dealership.

The number of service men needed varies with the work load. It is very difficult to operate a profitable shop with less than 65% labor recovery. This should be the basis for selecting the number of mechanics rather than by general guidelines. It must be remembered that the service department does more than just repair equipment. It creates confidence in the dealership, which in turn builds sales volume.

Good personnel policies demand that each man has an opportunity to rise to the top. A typical team in an average-size dealership will have a service manager, a field service man, three to four mechanics, a set up and delivery man, and a truck driver. Each should have the opportunity to move to a more skilled position. Vacancies should be filled from within whenever practical. When seeking a new employee, make certain that present employees know the requirements and are given the opportunity to recommend someone for the position. Typical job descriptions for the manpower needed in the service shop are shown in Appendix B.

EQUIPMENT NEEDS

The service department must be properly equipped to provide adequate service (Fig. 12.1). Mechanics will normally provide their own hand tools. All others must be provided by the dealership. These are capital investments that will improve service shop performance.

A properly equipped service department will contribute to:

Greater service department profit through increased productivity of servicemen
Greater employee productivity
Greater employee satisfaction
Greater ease in attracting and employing qualified service department personnel

FIG. 12.1. Shop doors should be sufficiently large to handle the largest equipment.

Safer service facilities for employees and customers
Greater customer confidence
Lower cost quality service
Improved levels of service quality and service performance with greater precision

Poorly equipped service departments experience:

Dissatisfied servicemen and high employee turnover
Low employee efficiency
Excessive rework of prior repairs and consequent loss of customer confidence
Excessive customer machine downtime
Excessive on-the-job injuries
Excessive expense in recruiting and training servicemen

In considering what tools to buy, primary consideration must be given to the necessity of a tool for completing a job. If the job cannot be done without the tool, the decision is obvious. If it can be done with reduced accuracy or with increased time, this must be taken into consideration. Tools should pay for themselves in labor saved or in improved job performance. Where tools are needed only infrequently, consideration should be given to sending out such jobs to specialty shops. A typical example is crankshaft grinding or cylinder head planing. If the service is not available in an area and is needed, then consider obtaining the tool or machine and selling the service in the area and adding to the service department income.

Many special tools are routinely used in the service shop. Special rates are not charged when these tools are used. Typical of these are valve seat refacing machines, drill press, hydraulic press, and spark plug cleaning and testing equipment. More expensive tools, requiring specialized skills to operate, will require special rates when they are used. Typical of the tools for whose use you must charge extra are the electric welder, cylinder boring machines, lathe wheel balancer, and paint spray machine.

To determine the extra charge needed to depreciate this equipment, determine the fixed cost per hour and the variable cost per hour. A typical computation may be as follows:

Electric Welding

Fixed cost:

Cost of welder	$1,000.00
Hours use per year	200 hours
Expected life	10 years

$$\text{Depreciation:} \frac{\$1,000}{10 \times 200} = \$0.50 \text{ per hour}$$

Variable cost:

Utilities	$0.25/hour
Material (rods)	$1.00/hour

Total cost (Fixed + Variable)	$1.75/hour

The shop is justified in marking up this service 100%. The cost for welding on this machine then should be $3.50 special charge per hour plus the conventional customer charge for service labor. If competitive rates in the area are lower than this figure, consideration should be given to reducing the charge to make it competitive yet still recover all costs.

Providing your shop with the necessary special tools will:

1. Increase customer satisfaction
2. Increase profits without increasing buildings or manpower
3. Attract better servicemen

Tools can be replaced from the profits as they wear out.

No attempt will be made to list special tools that should be purchased. This can be determined only on a local level.

The John Deere Company has this to say about special tools in their service management manual:

Service equipment classifications typically include:
A. Cleaning equipment.
B. Diagnostic and measuring equipment.
C. Lifting and holding equipment.
D. Disassembly and reassembly equipment.
E. Metal sizing and working equipment.
F. Routine service equipment.

Considerations concerning the selection and use of tools and equipment in each of these categories follow. The service manuals and technical manuals should be reviewed to determine specific needs and required capacities. Anticipated changes in power and equipment design, application and size should be considered. The number of service personnel in the shop and the likelihood for similar tool applications at any one time will affect the number of specific tools to purchase.

A. *Cleaning Equipment*—Appropriate cleaning equipment should be provided for shop clean-up and maintenance, for tractor and equipment assemblies and for sub-assemblies as well as individual parts. The availability and proper use of this cleaning equipment:

1. Allows thorough and accurate inspection and analysis.
2. Speeds disassembly and reassembly.
3. Promotes cleaner mechanics and cleaner work.
4. Improves customer satisfaction and confidence.
5. Requires less time to maintain a clean service facility.

Sufficient cleaning equipment is essential to the efficient and effective operation of a service department. Its use will contribute to improved service depart ment performance and increased service department morale.

Tractors and equipment that have been properly cleaned prior to diagnosis and repair will contribute to clean work habits and pride in completing the service operation.

An important point in considering cleaning equipment for complete assemblies or sub-assemblies is the convenience with which it can be started and used. A cleaning unit requiring little start-up time and with adequate capacity to quickly clean heavy dirt or grease from large units will be used frequently.

B. *Diagnostic and Measuring Equipment*—The availability and use of proper diagnostic and measuring equipment:

1. Reduces the cost of service.
2. Promotes more accurate diagnosis.
3. Promotes quicker repair time.
4. Results in less comeback service.
5. Allows accurate testing following completion of service operations.

Testing and diagnostic equipment is essential to profitable service department operation. The use of outside diagnostic services is impractical in many instances. Tractor and implement repair through "trial and error" parts replacement is expensive and seldom corrects the primary cause of failure. Diagnostic and test equipment quickly and accurately identify the cause of equipment malfunctions.

Diagnostic equipment should be purchased from a reputable supplier who can recommend manufacturers, models and sizes that effectively serve the needs of the service department. The Area Service Manager can also provide assistance in this area. The supplier should be able to provide initial and periodic instruction as well as quick repair service.

C. *Lifting and Holding Equipment*—Adequate quantities and capacities of lifting and holding equipment improve labor efficiency and working conditions and contribute to greater employee and customer safety. By providing a more ideal work environment, more precise and accurate work can be performed quickly. Equipment to minimize physical labor in the service department results in safer, more desirable working conditions. These conditions enhance the physical and mental alertness of employees, result in fewer time-consuming and costly injuries and assists in hiring and retaining well-qualified servicemen.

Supportive equipment for frequent jobs should be procured as soon as a need for safer, quicker and more accurate work is recognized. The investment will soon be returned in more productivity per hour, less comeback service, and greater customer satisfaction.

Equipment should have adequate capacity and versatility for current and projected service operations. Built-in safety features, fool-proof safety devices and overall safe design are of extreme importance. Portable lifting and holding equipment that can be conveniently maneuvered by one serviceman saves time and minimizes interruptions to other employees.

D. *Disassembly and Reassembly Equipment* (Fig. 12.2)—These types of tools also contribute to quicker and more accurate service. In many instances, they are essential in completing a job and are required regardless of the frequency with which the job occurs.

They should be relatively simple and convenient to use. Safety in moving, handling or supporting heavy assemblies is of prime importance. Time gained through rapid disassembly and reassembly will not detract from precision service if proper tools and procedures are used.

Special tool sets available from your supplier include these types of tools. These special tools are designed for specific operations.

E. *Metal Sizing and Working Equipment*—The amount of sophistication required in these types of tools is also dependent on the service objectives of the business. Certain tools and equipment are required due to their high frequency of use and relatively low cost of use. Examples would include: drills, grinders, saws, taps, cutters, hones, and reamers, etc.

Other tools and equipment with a high initial cost which are infrequently used may be impractical if similar services are available in the area. Examples are boring bars, milling machines, lathes, or crankshaft grinders.

F.   *Routine Service Equipment*—(1) Pick-up, delivery and field service vehicles will be required for service department and other dealership activities. Sizes will range from small pick-up trucks to large tractor-trailer combinations.

Small trucks will be required for picking up parts or components, for making deliveries of small items, for pulling small trailers and for occasional business transportation. A tilt-bed truck or other easily loaded vehicle for transporting tractors and equipment should be considered. A winch with capacity to load defective units is a must. This vehicle may also be used as a prime mover for an easily loaded trailer with multiple axles for large tractors or equipment. It may be practical to consider tractor–trailer combinations for large tractors or machines such as combines if the need for transporting these units over a great distance is experienced often.

Primary consideration should be given to the versatility of the particular fleet of transporting vehicles chosen. The units chosen should be flexible enough to be used effectively by the service, parts, and sales departments. Actual sizes should be consistent with current and expected machine dimensions and weights. Additional weight of mounted equipment should also be considered. Normal make, model and supplier considerations are dependent on local conditions and personal preference. Convenience, ease, and safety in loading and operating vehicles is essential.

Both transporting vehicles and field service vehicles should be equipped with two-way communications systems that contribute to more effective use of labor and equipment. Adequate capacity should exist to reach all areas of the current and proposed trade area.

Increased machine size and accompanying difficulty in transporting large machines leads to the probable performance of more service in the field in the future. Field service offers quick, "on-the-spot" repair with less lost time and less expense for pick-up and delivery. Availability of this service builds customer confidence and supports new and used equipment sales. The convenience of field service often results in field repair of minor problems which might otherwise develop into major machine failures.

Field service vehicles should be equipped with sufficient equipment to repair the common field failures expected. Typically, the field service vehicle would

FIG. 12.2. The engine stand takes the work out of engine overhaul.

contain: welding equipment, metal cutting equipment, vises, portable lift or hoist, chains and ropes, blocking and supportive devices, commonly used special tools, suitable containers for fluids, basic hydraulic testing and service equipment, basic diesel testing and service equipment, small supply of common repair parts, basic set of service and technical manuals, fire, safety and first-aid equipment, and a complete set of hand tools.

Lubrication Equipment—(2) A well-equipped lubrication center insures proper predelivery of new and used equipment. It contributes to thorough completion of customer service. Air-powered equipment for large jobs and portable hand operated equipment for smaller jobs are required. Adequate equipment presents a good image to the customer and encourages the improvement of customer lubrication equipment and lubrication habits.

Cleaning and Painting Equipment—(3) The investment in equipment for properly cleaning and painting used or repaired equipment is quickly recovered. Reconditioned used equipment that has a "like-new" appearance promotes a rapid turnover and assists in building profits in the sales, service and parts departments.

High pressure washers, degreasers and/or steam cleaners allow a professional job of painting to be done. The Area Service Manager and the local tool supplier can provide information on specific features and limitations of cleaning equipment. Consideration should be given to cost of purchase, cost of use, cost of materials consumed, start-up time required, amount of water used, water pressure required, cleaning limitations and operator safety.

A painting area separate from the moisture-laden cleaning area will contribute to a good job of painting. A high quality paint gun adaptable to a self-contained one to two quart container or to a bulk container should be considered. Adequate housing to move around the largest piece of equipment and oil and water inline extractors are also required. Availability of portable drying lamps will contribute to high quality work.

An exhaust outlet near the floor to remove heavy fumes and over spray is required for convenience and safety. A dust-free inlet contributes to the desirable appearance required for rapid movement of used equipment. In cold climates, it may be necessary to heat incoming air to permit continuous painting in cold weather.

Danger of explosions in painting areas requires explosion-proof electrical service and periodic safety inspections. Respirators and other personal safety equipment must be available and used by employees. They provide a safe work environment in addition to improving employee satisfaction and performance.

An air compressor with sufficient volume to support paint guns, impact wrenches and lubrication equipment without overloading the unit should be available. Separate painting and service buildings may require an air compressor in each area.

State and local zoning, insurance, safety, and fire regulations for painting facilities should be checked to insure compliance. (See OSHA Standard Sec. 1910–94 and 1910–107, "The Federal Register.")

Summary—The decision on tool and equipment purchases must be made in view of the present and anticipated service needs and the availability of specialty services and supplies.

Service manuals and technical manuals should be analyzed to identify jobs which frequently occur, jobs requiring special equipment, and jobs that could be completed more conveniently with special equipment.

The alternative cost of contracting special service jobs out should be compared with projected costs of ownership and use.

Based on this type of an analysis, special tool and equipment needs can be established. Supplier reputation and available services should also be considered.

Tool maintenance and repair services should be available to keep tools and equipment in good working order. Periodic instruction on sophisticated equipment for all servicemen should be encouraged.

The essential and convenience features of tools and equipment must be matched against their need in service operations. Tools and equipment to facilitate service operations contribute to improved individual serviceman and service department performance.

Tools and equipment purchased but not used or improperly used adversely affect service department profits.

Equipment that is needed but not available adversely affects labor efficiency, service department performance, and the service image.

OPERATING AT A PROFIT

The single most important contributor to a profitable service operation is labor recovery. You must balance labor purchased with labor sold. Your recovery of labor must be at least 60% and you should preferably strive to obtain 75%. To do this, let's first look at the shop ticket. What do we cost to the ticket?

All labor
All parts
All supplies
All travel

Tickets must be written and time recovered for all shop activities, which include:

Repair customer's equipment
Assemble, set up, and inspect new equipment
Repair and recondition used equipment
Delivery and instructing owner of new equipment
All other internal work done for the dealership such as repairs of
trucks and shop equipment

Shop tickets must be handled so that none are lost. Have them numbered consecutively. Account for all numbers. The procedure listed here is a good one:

1. All tickets should be made up by the shop foreman.
2. Make tickets in triplicate.
3. The hard copy is for the mechanic; it is also his basis for obtaining parts.
4. The mechanic records time spent on each ticket.
5. The foreman or service manager approves the completed ticket and turns them in to the bookkeeper.
6. The bookkeeper extends ticket.
7. The original ticket is mailed to the customer, the second copy is retained on file by the bookkeeper, and the hard copy is retained by the service department for their records.

In a typical dealership, internal work often exceeds customer work. It is difficult to assess the proper mix for any area. It may vary from 40% customer to 60% customer work. The correct mix must be arrived at according to local conditions. More important than the mix is productivity, as shown by the labor recovery rate.

Labor recovery rate is the ratio of hours sold to the hours of labor purchased.

$$\text{Labor recovery rate, as percentage} = \frac{\text{Hours sold} \times 100}{\text{Hours purchased}}$$

Because of holidays, sick leave, vacation, coffee breaks, jury duty, and the like, it is never possible to recover 100%. In fact, 82.5% is considered maximum, and 70% is very good. Since all labor purchased cannot be sold, pricing labor becomes very important.

Some Formulas to Apply

First, let's consider the break-even point for a shop. This is the dollars required to meet labor payroll and overhead.

$$\text{Break-even } \$ = (\text{Hours purchased} \times \text{Wage rate}) + \text{Overhead } \$$$

The overhead is the total dollar value of overhead expenses that must be recovered by the service shop.

Next, let's compute an hourly rate to charge customers. This must be based on an anticipated realistic recovery rate.

$$\text{Hourly rate} = \frac{\text{Break-even} \div (1 - \% \text{ profit})}{\text{Paid time} \times \text{Recovery rate}}$$

Consider a shop using eight mechanics and one service supervisor who does no ticketed work. The shop purchases 1280 hours (160×8) per month and pays mechanics \$5.00 per hour. The shop supervisor draws a straight salary of \$1500 per month. Additional overhead charged to the shop amounts to \$2400 per month. Seventy percent of the labor is recovered. A profit of 10% on sales is desired.

$$\text{Break-even} = (1280 \times 5) + \$1500 + \$2400 = \$10,300$$

$$\text{Hourly rate} = \frac{10,300 \div (1 - 0.10)}{1280 \times 0.70} = \frac{11,444}{896} = 12.77$$

Round this figure out to \$13.00 per hour.

Potential labor sales for the month can now be computed:

$$\begin{aligned}
\text{Potential labor sales} &= \text{Paid time} \times \text{Recovery rate} \times \text{Customer rate} \\
&= 1280 \times 0.70 \times 13.00 \\
&= \$11,648
\end{aligned}$$

On this basis the net profit for the shop per month will be:

$$\text{Net profit} = \$11,648 - 10,300 = \$1,348$$

$$\text{Net profit} \% \text{ of sales} = \frac{1,348}{11,648} = 11.57\%$$

The example shows that the customer's rate must be set at 2.5 to 3 times the wage rate. It further points out that overhead must be kept down if customer rates are not to get out of hand.

Examine what happens if the labor recovery rate drops by 10% down to 60%. Our potential sales are now:

$$\begin{aligned}
\text{Potential Sales} &= \text{Paid time} \times \text{Recovery rate} \times \text{Customer rate} \\
&= 1280 \times 0.60 \times 13.00 \\
&= \$9984
\end{aligned}$$

As our break-even is still \$10,300, we now have a net loss.

$$\text{Net loss} = \$9984 - 10,300 = \$316$$

At \$13.00 an hour for labor, our lowest break-even recovery rate can be obtained by equating the potential sales to the break-even point and solving for the recovery rate:

$$\text{Break-even} = \text{Potential Sales}$$

$$\$10,300 = \text{Paid time} \times \text{Recovery rate} \times \text{Customer rate}$$

$$\text{Recovery rate} = \frac{10,300}{1280 \times 13.00} = \frac{10,300}{16,640} = 0.619 = 61.9\%$$

The importance of labor recovery rate can be seen from the break-even analysis shown for the eight-man shop for one month in Table 12.1.

TABLE 12.1. BREAK—EVEN ANALYSIS FOR EIGHT—MAN SHOP

Labor rate	Hours[1] sold	Recovery[1] rate (%)	Total revenue	Total[2] cost	Net profit ($)	Net profit (%)
13.00	704	55	9,152.30	10,300	(1,147.30)	(12.5)
13.00	768	60	9,985.54	10,300	(314.46)	(3.1)
13.00	792	61.9	10,300.00	10,300	0.00	0.0
13.00	832	65	10,818.50	10,300	518.50	4.8
13.00	896	70	11,651.58	10,300	1,351.58	11.6
13.00	960	75	12,484.80	10,300	2,184.80	17.2

[1]Hours purchased = 160 × 8 = 1280 hours.
[2]Total cost = cost of labor + overhead.

CONTROLS

Controls cannot substitute for management, but there are no substitutions for controls in management. Since the service department exists to sell time, it follows that the heart of service shop management is controlling time.

Control starts with the shop supervisor. He has the prime responsibility for the efficient scheduling of work. He should maintain a weekly forecast of scheduled work. This enables him to balance the workload, to secure the necessary parts for scheduled work, and to minimize the time that customer equipment is laid up. In addition, jobs requiring special skills can be scheduled when the appropriate mechanic is not otherwise engaged.

The daily schedule is also the responsibility of the shop supervisor. Several forms are available for this purpose. Use one that has a column for "Estimated Time" and "Actual Time." It is of great help in determining who needs additional training or job counseling. The shop supervisor should have work orders ready for assignment as work is completed. Low labor recovery rates can often be traced to poor scheduling techniques.

Require each mechanic to post his time to the shop ticket. National Farm and Power Equipment Dealers Form FER 258 is a convenient form to use. It will show you how much time was expended on each job order. The axiom "The men do well what the boss looks after" is very applicable here. Look at the individual labor recovery rate for each mechanic weekly (Fig. 12.3). Go over it with him. It is surprising the effect that has on a mechanic. Don't be surprised if the reaction is negative at first. Show him how his recovery rate affects the labor recovery rate goal for the overall department. In the end, discussion never fails to motivate the kind of mechanic you want. Additionally, such a record is a necessary document when salary schedules are discussed or incentive pay plans are being considered.

Play fair with your mechanics. Credit them with all the internal labor they are assigned. The obvious internal work on trade-ins is never a problem, but are you crediting them with time spent at

The form in the figure reads:

Week ending ____________   Individual____________

SUMMARY

I. Hours Paid

    Weekly Hours Paid   ______hrs.

II. Hours of Productive Work

	Hours Sold	Hourly Rate	Gross Earnings
A. Customer			
1.______________	_____	$______	$______
2.______________	_____	_____	_____
3.______________	_____	_____	_____
4.______________	_____	_____	_____
5.______________	_____	_____	_____
Total Customers	_____	_____	_____
B. Internal			
Assembly and Handling			
1.______________	_____	$______	$______
2.______________	_____	_____	_____
3.______________	_____	_____	_____
Recondition Used and Other Internal			
1.______________	_____	$______	$______
2.______________	_____	_____	_____
3.______________	_____	_____	_____
Total Internal	_____	_____	_____
Total Productive Hours Sold to Customers and Internal	_____ Hrs.		

C. Ratio of Hours Paid divided into Hours of Productive Work =

                                        ______%

FIG. 12.3. Work sheet for individual labor recovery.

demonstrations, building exhibits, building maintenance, equipment maintenance, and work performed in the parts department? There will always be some unassigned time due to paid time off, lack of work, and scheduling inefficiency. The shop supervisor must take the responsibility for keeping the latter two to a minimum.

Warranty labor is always a problem. The service manager should review all warranty work. Care must be taken to process all claims and to verify that they are settled. Follow the procedure called for by the parent company. Replaced parts usually must be secured and held for inspection.

Follow up on reworks by mechanics. They may point to potential problems. The time clock is the best means of recording on and off shop tickets. "No charge" shop tickets must be reviewed by management. This is an easy way for mechanics to cover up "mistakes." It's also an easy way for service managers to satisfy irate customers.

The effective service manager uses control charts (Figs. 12.4 and 12.5) to guide the service department. The charts should establish a break-even point and set a suitable goal. It can be set in terms of margins, gross sales, or labor recovery rate.

Charts of this type identify trouble before it gets out of hand. It makes management at all levels aware of what is happening. This is important. The most important aspect, however, is to take the necessary corrective action.

INCREASING SALES

When sales are down one of two things must be done—reduce labor purchased or increase labor sales. Reducing labor purchased is a serious matter that should be considered carefully before it is done. This alternative should be the last resort. If more than half of the months are below the break-even point, consider reducing labor purchased.

Every effort should be made to increase labor sales. If you have been charting, lean months should be obvious. Plan ahead to overcome those conditions. Some of the following ideas will work for you as well as they have worked for others.

Advertise.   Run ads in your community newspaper. Spot announcements on radio and TV are effective. Send out postcards to selected customers or include handbills with the monthly statements. Have everyone in the dealership conscious of the value of advertising by word of mouth. Use displays near the sales department or in show windows.

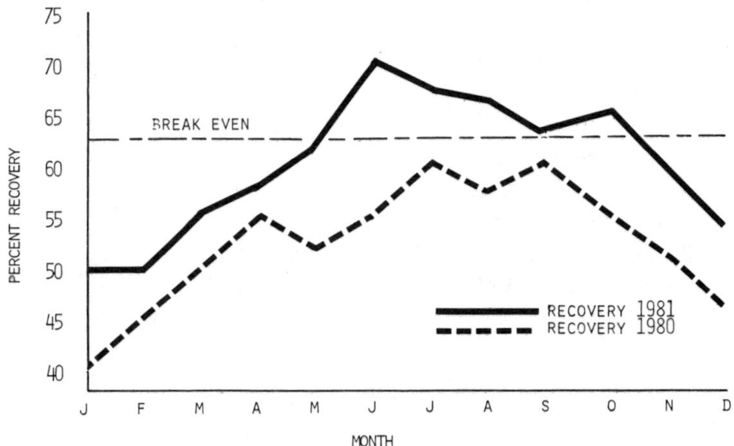

FIG. 12.4. Labor recovery control chart.

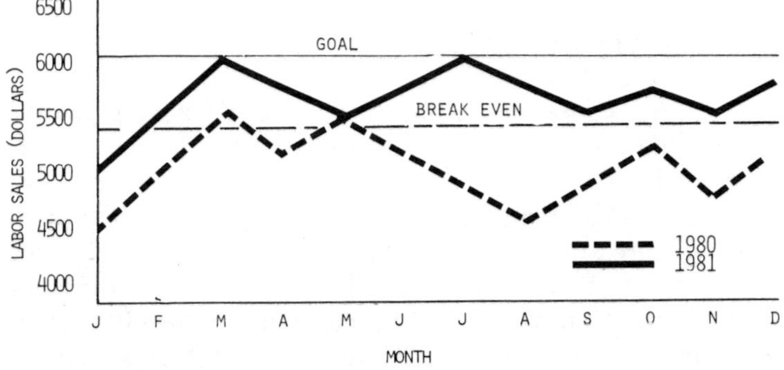

FIG. 12.5. Labor sales control chart.

Solicit.   The service manager should know his customers. Have him call customers who he feels may need service work. Offer an incentive to servicemen who schedule repair jobs.

Offer specials.   Tractor service specials are an ideal way to stimulate sales. Tractor repairs are usually high-volume jobs that promote service labor and spare parts. Price out the typical overhaul job for the popular tractor in your area. Feature this during the lean months. You may want to consider offering an "overhaul now, pay later" proposition. Limit the time when payment will become due. Make arrangements with the bank to carry the additional receivables. You may want to take a chattel mortgage on the tractor and discount the notes at the bank so that you may use your funds for other operations. Of course, you pay the interest for the stated time.

The dynamometer check can also be a popular program. Customers like to know they are getting maximum fuel efficiency today. This check results in tune-up jobs as well as major overhauls.

Personal salesmanship on the part of the service department personnel must be combined with publicity to do the job. The following basic principles should produce good results:

1.   *Make customers like you.* Conduct your business in such a way that people will want to bring their mechanical service problems to you. You can do this only with friendliness, a spirit of cooperation, and high technical skill.

2.   *Be always helpful.* Always give the type of service you would like to receive. Advise the customer what he needs. See that the work is done properly. Explain why you diagnosed his problem as you have. Let him know the consequences if repairs are not made.

3.   *Help the customer decide.* Most customers want to operate as economically as possible. Help them determine how to do this.

4.   *The customer is always right.* If he is not, it will not help things to argue the point with him. Agree when you can. When you cannot, use the "yes, but" technique. *Yes,* you can get it cheaper at the crossroads shop, *but* not with a 90-day guarantee.

5.   *Be appreciative.* Let the customer know that you value his business. He is a preferred customer. A thank you for your business is always in order.

6.   *Give the customer a little extra.* The French have an expression for this, "lagniappe." Try it. If it is nothing more than a battery check, write it up with an N.C. (No Charge). Everybody likes something for nothing.

7.   *Know your equipment.* This is basic. You have to keep posted on the latest, or you will soon be a "has been."

8.   *Don't knock competition.* This cannot help. Stress your strong points. The customer knows your competition's weak points or he would not be at your shop.

9.   *Use the telephone.* When not busy, talk to your customers. Help them decide on the best time to get repairs done. This gains their respect and good will.

10.   *Avoid misunderstandings.* Agree on what is to be done with the customer before starting repairs. If changes have to be made, contact the customer. Sometimes this will not be possible. If the work runs over the estimate, have the bookkeeper send you the bill so that, when the customer comes in for his equipment, you can personally explain the changes to him.

11. *Inspect every job before delivery*. Make sure the customer gets what he paid for. This makes satisfied customers and holds down after-sales expenses.

12. *Utilize the human senses to sell*. Show him what you propose to do. Tell him what you propose to do. Let him feel the roughness of the cylinder bore. The decision to buy is a mental one. Use everything you can to stimulate that mind.

Finally, keep your service area clean and attractive. Assign areas to everyone. Clean up in the evening before leaving.

QUESTIONS

1. How does selling labor differ from selling other commodities?
2. What are the primary duties of the service supervisor?
3. How would you determine the extra charge for the use of a specialized piece of equipment?
4. How would you determine what specialized tools to buy?
5. What items of cost should be on every shop ticket?
6. What shop activities would you include on a shop ticket?
7. What is a "break-even point?"
8. How would you compute labor recovery rate?
9. What is a "control chart?"
10. How might you stimulate labor sales during slow periods?

13

Personnel Management

Good personnel management requires leadership. General Dwight D. Eisenhower defined leadership as the ability to get others to do the things you want them to do because they want to do them. Your employees are an extension of you. Getting them to react to situations as you would react is what it is all about.

It is difficult to get good leadership without discipline. Discipline is a much-maligned and often misunderstood word. It may be equated to the autocratic discipline demanded of soldiers by the Prussian Army. Certainly, there is no place for this in American business.

Discipline is nothing more than a conditioned reaction to the situations that surround us. The wise leader conditions himself to his surroundings and in turn conditions his team to the conditions they will face. Making others see the necessity for this and accepting it cheerfully is the job of the leader.

The policy that states, "The customer is always right," is a sound one but not necessarily a correct one. The disciplined employee will answer the customer's complaint with a "Yes, our mechanics are not fast but their work is the very best." The undisciplined employee might well answer the same question with a "Take your junk home and see if you can repair it any faster." Management has no job more difficult or more important than conditioning employees to react as they would.

COMPENSATION

The single most important factor in good employee relations is pay. "Other benefits," particularly with your higher-paid employees, is a category rapidly advancing in importance. Adequate pay for employees is rewarding both to employee and employer. With a sound balanced pay policy we usually end up with a situation featuring:

Good, well-qualified people seeking employment
Present employees staying with the firm
Less productive employees challenged to improve or find other employment
Harmonious working relations
Management with more time to manage

This never happens without a clearcut policy on wages. Time and wages must be balanced—the employee's time against the employer's wages and fringe benefits. This equation must be kept in balance at all times. The policy must include most if not all of the following:

Holidays	Incentives
Vacations	Pensions
Sick leave	Health insurance
Compensatory time	Life insurance
Overtime pay	Vesting rights

Good relations result only when both feel benefited.

Wages must be competitive in the area. The location, size, and social level of the town affects wage rates. The working conditions and hours must be taken into consideration.

Wages are not necessarily the best way to encourage maximum performance. Management must consider profit sharing, bonuses, and incentive pay plans as well. The purpose of any plan is to stimulate employees to do their best. Any good pay plan can be easily evaluated. It must:

Stimulate each employee
Stimulate them often
Benefit both employee and employer

Wages or salaries fall short when judged by such criteria. Good wages are soon accepted. Their value as a stimulant is soon forgotten. Unless the employee increases production with each pay raise, it cannot be shown that the employer benefits.

The profit-sharing plan has merit. A certain percentage of the profit is set aside and divided among employees. As it is tied to profits produced by the company, from that standpoint it is sound. It requires careful bookkeeping. It should be paid at least quarterly to stimulate as often as possible. A certain percentage can be held back to cover slack seasons and a final accounting made at the end of the year. The profit-sharing plan does have two shortcomings:

It fails to credit the individual worker.
It does not stimulate employees often enough.

The bonus is used by many firms. It is usually set up on a variable basis. The employee seldom knows exactly what it will be. This plan also has shortcomings.

It may not credit the individual.
It does not stimulate often enough.
It often becomes expected rather than earned.

The incentive pay plan combines the advantage of both plans without their apparent disadvantages. As it is an individual thing, it takes many forms. It is based on a base pay of a prescribed amount plus a monthly or quarterly amount. Set up properly, it appears to be the most effective.

TYPICAL INCENTIVE PLANS

Service Center

Mechanics can be employed according to two different incentive plans. A popular way is to pay a small guaranteed salary and a fixed percentage of the labor sales they make. Together this should not be more than 45–55% of the labor sales. The mechanic should be required to do comeback work on his own time. This plan may be in conjunction with the flat rate system.

In the flat rate system, each shop job is described in a master book and is assigned a certain time for completion. The customer pays a fixed amount for labor based on a set customer labor rate. The time allowed is based on the average time required for the job by the average mechanic. The mechanic gets 45–55% of his labor sales. The good mechanic will do the job in less time. Mechanics are encouraged to work faster. Both employee and employer may benefit by this system. Management must be careful to protect the rights of their customers, for this system can encourage the mechanic to do slipshod work in order to "gain time."

The service center supervisor has to be paid on a different basis. Consider a flat salary based on operating the service shop at a break-even point. He is then paid an incentive of up to 5% of labor sales above the break-even point. One-half of the amount should be paid monthly with a final accounting at the end of the year. In this way, both management and the employee profit when the service center makes money.

Parts Department

During the year 1980 average sales of parts per parts man, according to the Cost of Doing Business Study, ranged from $138,510 for the under $1,000,000 annual sales dealer to $172,037 for the $1,000,000 to $2,000,000 annual sales dealer to $211,895 for the over $3,000,000 annual sales dealer. With this in mind, let's consider an incentive plan for the partsman.

Again a flat salary might be paid either based on a break-even operation or based on the current volume of business if the plan is being added to an existing business. This salary should be based on up to 30% of the margin for that volume of business. An incentive of 5% of sales over this volume may be used.

Let's look at this example. Mike operates the parts department alone. Last year he sold $145,000 of parts with a margin of $42,000. He was paid $1000 per month. Additionally, parts turnover was 2.1% and obsolescence was 0.5%. There is room for growth in this new agency.

An incentive program could be set up in the following manner. Mike is placed a a flat monthly salary of $1000 per month with an incentive pay of 5% of sales over $12,000 per month. Pay Mike 2.5% every month and at the end of the year pay in lump sum whatever amount is needed to bring his incentive pay to 5% of everything over $145,000 for the year. Certain safeguards must be written into the plan. Obsolescence must not exceed 0.5%, turnover must be maintained over 2.0%, and a penalty must be applied if margins drop.

For that reason the incentive can be tied to the margin rather than to sales. A figure of 25% of the margin above $42,000 per year can be used. Again, it should be given on a monthly basis with a final accounting at the end of the year.

The incentive program has to fit the individual dealer. Consider the case of the dealer who has reached the point where there is too much work for one man and not enough for two. Perhaps computerized parts control is the answer. This will increase overhead in the parts department. In this case the incentive program may be tied to a break-even point rather than previous sales.

Pay the partsman his existing salary for continuing his current sales effort. Compute the break-even point for the department considering the additional overhead anticipated due to computerized parts inventory control. Pay a percentage of the margin over that necessary to cover the break-even point.

As an example, the partsman has been making $800 per month and has been selling $100,000 per year with a 30% margin. Additional overhead for the department has been 8% of sales and should decline with computerized parts control and increased volume of sales.

Present break-even point

$$\text{Labor} = 800 \times 12 = \$9600 \text{ per year}$$
$$\text{Additional overhead} = 100,000 \times 0.08 = \$8000$$
$$\text{Break-even} = \$17,600$$

Mike would then continue to receive $800 per month and a percentage (up to 25%) of the annual increase in margin. The company now has the following net profit:

$$\text{Gross margin} - \text{break-even} = \text{net profit}$$
$$\$30,000 - \$17,600 = \$12,400$$

Under the incentive sales program, sales can be expected to rise to $130,000 and margins be maintained at 30%.

Mike now draws incentive pay of:

$$0.25 \ (130,000 \times 0.30 - 100,000 \times 0.30) \text{ or}$$
$$0.25 \ (\ 39,000 - 30,000) = \$2250.00.$$

His total pay is now $9600 + $2250 = $11,850.

For an incentive plan to be good, the business must also profit. Let's test this. Net profit is now:

$$\text{Gross margin} - \text{break-even} = \text{net profit}$$
$$\$39,000 - [(130,000 \times 0.08) + \$11,850] = \$16,750$$

As a percent of sales in the parts department, the net profit is now $16,750 ÷ $130,000 = 12.9%. Before the incentive plan went into effect the net profit was $12,400 ÷ $100,000 = 12.4%. The profit picture is improved and the employee is now definitely profit oriented.

Figures and percentages shown here are for illustrative purposes only and not intended to be used as an infallible guide.

Sales Department

During the year 1980, the average sale per salesman ranged from $267,716 for the under $1,000,000 annual sales dealer to $428,064 for the $1,000,000 to $2,000,000 dealer, to $542,757 for the $2,000,000 to $3,000,000 dealer and $734,841 for the over $3,000,000 annual sales dealer. This wide variation in sales potential complicates the incentive plan. Remember that the salary, commission, and so on plus expenses of the salesperson should be held to 5% of sales. When you consider that margins on new and used equipment for all dealers average only 10.08% and 7.05% respectively, the importance of controlling selling expenses becomes very important.

An incentive plan for salesmen can be designed around a salary and commission. For the first $300,000 of sales, pay a salary of 4% of sales or $12,000 per year. One percent is held for the salesman's expenses and delivery. This is based on maintaining a margin of 12% of sales. The salesman can be given 4% of all sales over $300,000 per year. Incentive should be paid on a monthly basis. Management must reserve the right to approve all trade-ins. Definite ground rules must be set as to what constitutes a sale.

One management consultant recommends that total compensation to the salesman not exceed 25% of the first $24,000 of sales margin developed plus 35% of all over $24,000. The advantage of tying the plan to margin rather than to sales is evident. Many will question the desirability of increasing the sales cost as sales go up.

A firm selling between $1,000,000 and $2,000,000 annually normally employs two salespersons. Typically, each will sell from $400,000 to $450,000 of equipment and develop margins of 10–12%. Let us develop a compensation plan for the salesmen who sell $425,000 of equipment and develops $46,750 of margins or 11% of sales.

Based on sales, the earnings of the salesperson would be:

$$\$300,000 \times 0.04 = \$12,000 \text{ salary}$$
$$\$125,000 \times 0.04 = \$\ 5,000 \text{ incentive}$$
$$\text{Total pay} = \$17,000$$

Using the criteria based on margins the salesperson would earn the following:

$$\text{Total margin} = 0.25 \times \$24,000 = \quad \$6,000.00$$
$$0.35\,(46,750 - 24,000) \qquad = \quad \underline{\$7,962.50}$$
$$\$13,962.50$$

It is evident that the latter plan will encourage the salesmen to develop healthy margins whereas the first plan can result in sales without profit to the owners. It is never easy nor desirable to reduce a salesman's pay. When shifting from straight salary to salary plus commission, let the salary guarantee remain the same. The sales base should be such that the salary does not exceed 4% of sales. Any commission paid should be for increased sales. Unless margins are maintained and used equipment is sensibly appraised and sold, management can find itself working for the salesman. It must be stressed that accurate records must be kept and carefully thought-out rules must be spelled out as to who gets credit for sales participated in by more than one salesman.

PROFIT SHARING

The profit-sharing plan has worked well in some industries. It is based on dividing a certain percentage of the profit among all employees. It tends to motivate all employees to help in recruiting good workers. It tends to encourage all workers to see that everybody in the organization produces.

It does have faults. It fails to credit the individual for his effort. Unless an unusual amount of work is done by the bookkeeping staff, the profit cannot be distributed often enough to keep employees stimulated.

Salesmen's pay plans vary from straight salary to percentage of sales, percentage of margin, or percentage of company profit, to stock options, special fringe benefits, and combinations of the above. Choose a plan that will motivate often and profit the employee and employers, but keep it simple. Elaborate plans generate a bookkeeping load and a great potential for misunderstandings.

Management, to repeat, is the unifying force that coordinates resources and directs action. The definition applies to personnel management as well. It is well to understand the differences among management, supervision, and operation. The good manager divides and balances his time between day-to-day operational problems and the essential thought, analysis, and planning he must do. He delegates authority and responsibility so that he has time to manage as well as to develop leadership among his subordinates.

Management is divided into three categories—planning, operating, and supervising. To illustrate, the manager is planning when he makes the profit control plan for the company. He is operating when he sets up a shop labor recovery chart for the service department. He is supervising when he accompanies a new salesman on his sales rounds.

The mix of operations, supervision, and management that the manager performs is a function of the size of the organization and the number of assistants available to help. In a large firm it may be practically all pure management. As the level of management goes down, the amount of supervision increases in comparison with pure management. The management triangle may serve to clarify (Fig. 13.1). It rests on a sound base of supervision. The manager of a typical retail store discharges the functions of top and middle management.

The manager of the local implement store must of necessity wear many hats. Nothing is wrong with that. What must be avoided is the temptation to try to do it all. Evaluate the skills of your employees and delegate work. Delegate all the jobs that might be done by others as well as you could do it. Learn to delegate jobs that may not be done quite as well as you would do it but that will serve to train and develop your employees.

Delegate jobs that you do not have time to accomplish even though they might not be done quite as well as you would do them. Give yourself ample time to do the important thing—manage.

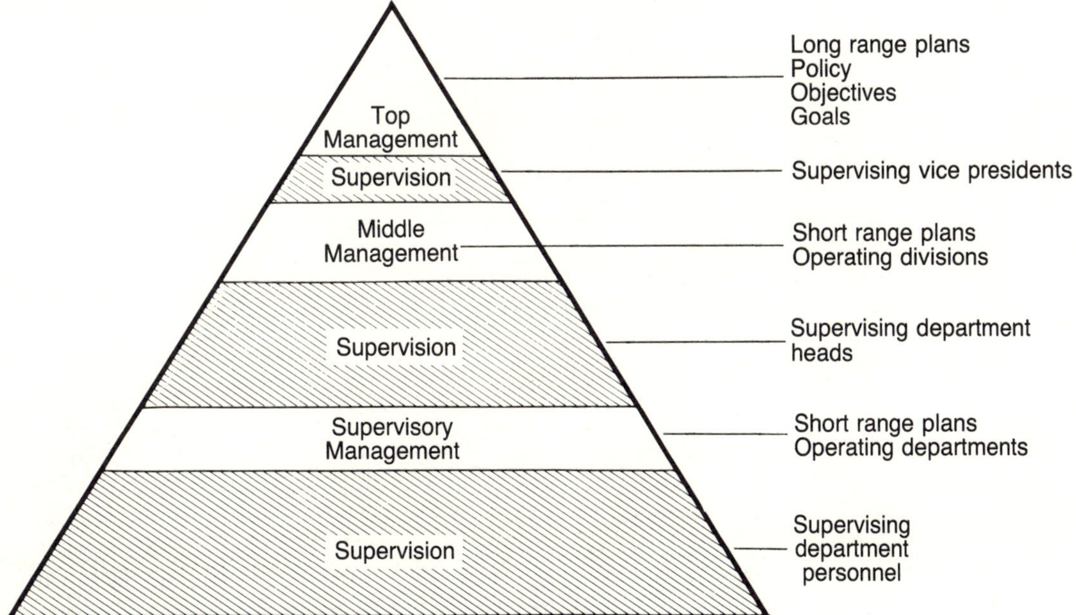

FIG. 13.1. The management triangle.

Functions of Management

1. Management must do the planning for the organization. It must set the goals. It must sit back and think, judge, evaluate, and decide the where, the when, and the how. It must determine alternatives to guide along uncharted paths.

2. Management must be the organizing force. It must first build the organization, then structure it to meet the needs. It must assign supervisors, group skills, and fit people into the work.

3. Management must direct. Decisions are given form by direction. Capital must be directed to profitable areas; resources must be directed to areas that are in need.

4. The manager must coordinate. No department is an island that can stand alone. Parts supply must be coordinated with service. Repair of used merchandise must be coordinated with used sales. Assembly of new equipment must be coordinated with whole goods. Accounting must be kept attuned to all departments. Most important, personnel must be coordinated. It is a constant job of combining people, material, equipment, and facilities to accomplish the distribution of parts, service, and equipment.

5. Management must review and audit. Plans must be kept updated to changing conditions. Accomplishments must be measured. New decisions must be made to keep the organization headed in the desired direction.

Perhaps this definition of management is an ideal summary of the functions of management. Management is the science or art of combining ideas, facilitating processes, materials and people to produce and market a product or service profitably.

Functions of Supervisors

The supervisor is concerned with people rather than machines. He deals with the present rather than the future. He sees that plans are fulfilled and decisions are carried out. His activities are as follows:

Assigning work
Reviewing results
Instructing employees
Checking schedules
Obtaining supplies and material
Devising work improvement methods
Coordinating jobs
Handling technical matter

To summarize, the manager's job is concerned with the future, and the supervisor's job is concerned with the present.

SELECTING PERSONNEL

A business can be no better than the people who man it. For that reason, no job is more important than selecting the right employees. Two methods are normally used, the human approach and the job approach.

In the human approach, you attempt to find and hire a top-notch man, then fit him into the position. There is merit in this, to a point. The conscientious, willing, intelligent individual can be trained for many jobs. This works if you have time and people to do the training. It has limited value for specialized skills such as mechanics.

TABLE 13.1. JOB DESCRIPTION: PARTS MAN FOR HIGH-PROFIT IMPLEMENT COMPANY

Approximate share of time (%)	Duty	Responsibilities
50	Stock maintenance	Maintain perpetual inventory as well as take periodic inventories. Maintain and assist in analyzing records of stock movement, by season and year. Order stock.[1] Receive stock (check shipments against orders and invoices for shortages or overages). Organize stock room for most efficient operations. Stock shelves and bins. Maintain file on local sources of parts for own and competitive equipment. Keep posted on new equipment and parts (by file and personal memory).
35	Selling	Know to whom credit may be extended. Know discounts for wholesale parts. Make tickets for all sales. Maintain order sheets for repair jobs. Keep list prices up-to-date.
10	Assistance to other departments	Assist bookkeeper with questions on parts billing, invoices, discounts, etc. Assist management, sales and service with questions concerning credit ratings of customers. Assist management and sales with customers' questions concerning frequency of repairs on various items of equipment.
5	Reporting	Prepare and present such reports relative to the operation of the department as immediate supervisor requests.
100	Public relations	Maintain a cheerful, helpful attitude with employees and customers. Maintain a neat, orderly work area. Serve customers promptly and rapidly. Know thoroughly and explain company policies and procedures when employees or customers raise questions concerning them.

Adapted from "Third Annual Conference for Farm Equipment Dealers," Blacksburg, Virginia, 1962.
[1]Up to $2500 month. Over this amount requires the approval of immediate supervisor.

The job approach is perhaps better suited to the retail implement and tractor dealer. Start with a job analysis. What will this employee be required to do? What skills must he possess? What are his responsibilities? What authority will he need to do his job? Are there any special educational needs?

With this in mind, go to the job description. It should be written and available to use during the interview. The job description is a listing of the responsibilities of the job, the authority needed to discharge the responsibilities, and a breakdown by time devoted to the various activities the job envisions. A typical job description for a parts man is reproduced in Table 13.1. Writing these takes time but is well worth the effort. It serves as a guide for the employee and is an invaluable tool for supervisors when it is time for performance evaluation. Typical job descriptions for service personnel are shown in Appendix B. Less detail is shown on those forms.

Next consider personal requirements. Here you consider the individual characteristics of the man. Again, it should be reduced to writing. For the parts man, these requirements might be as shown in Table 13.2.

TABLE 13.2. PERSONAL REQUIREMENTS AND CHARACTERISTICS OF PARTS MAN FOR HIGH-PROFIT IMPLEMENT COMPANY

Mental ability and attitude:
    Cheerful
    Wants to help people
    Pays attention to details
    Good organization ability
    Good with figures
    Conscientious
    Intelligent
    Eager to learn
Appearance and physical traits
    Clean
    Neat
    Good eyesight
Previous training and experience
    High school education (upper half of graduating class).
    Some familiarity with machinery (names of parts and
      general function of them).
    Course in business arithmetic desirable.

Adapted from "Third Annual Conference on Business Management for Farm Equipment Dealers," V.P.I., Blacksburg, Virginia, 1962.

Armed with the two documents outlined in these tables, personnel selection can begin. Use a job application form. Remember that some things can be discussed now that will never seem proper in the future. It is very easy to turn down a potential employee but much more difficult to discharge an employee. Mr. C. W. Whitney, Executive Director of the Farm and Power Equipment Retailers Association of Ohio, suggests the following checklist to help gather information on which to base an opinion before hiring.

1. Why does the applicant want to work for you? Ask him. Maybe he wants a better job than he now has. Is he out of work, and why?
2. Do dates of past jobs show lapses in employment? Why? Has he been unemployed? Did he draw Unemployment Compensation?
3 Does he make courteous remarks about past employers? Does he speak a good word for other people with whom he has worked? Does his personality seem to fit the pattern of your present staff?
4. Why did he leave his last job? How much notice did he give the last employer? How long does he plan to work for you? Does he have a record of staying with a good job'
5. What kind of work has he done previously? Is his experience of value to you? Has he progressed on past jobs as indicated by salary increases and responsibilities handle
6. Has he been happy on past jobs?
7. Has the applicant ever been in trouble? Arrested?
8. What is his financial background? Does he have a good credit rating? Is his salary garnished?

With such information in hand, take time to consider carefully each applicant. Upgrade personnel by replacing poor workers with good men and women, then plan personnel relations so that they will want to stay.

Application for Employment (Fig. 13.2)

Many questions are easier to ask in writing than they are in person. Much can be learned about a person by the way he expresses himself in writing. Does he have a handwriting that is legible? Does he use good grammar in expressing himself? These points should be observed in view of the type of work the person will do. The sample application form shown on the following pages could be used to help find better people.

FIG. 13.2. Employment application form.

APPLICATION FOR EMPLOYMENT

Date _______________________________

Name ____ Social Security No. _______________________________
(Print) Last Name, First, Middle

Present
Address ____ How long have you resided there? _______________________________

Previous
Address ____ How long have you resided there? _______________________________

Home
Phone _______________________ If you have no home phone, how can you be reached? _______________________________

Position Desired _______________________________________ Are you a citizen of this country? _______________________ State age if under 17 or over 70 _______________

EDUCATION (Circle year completed)

Elementary School 1 2 3 4 5 6 7 8   High School 1 2 3 4   College 1 2 3 4 5 6

School
Attended _______________________________   _______________________________   _______________________________

Have you ever been bonded? _______________   Do you own a car? _______________   Do you have a valid operator's license? _______________   State, date issued _______________

Who, if anyone, recommended you to us? ____

MILITARY SERVICE:

Date Entered _______________   Date Discharged _______________   Branch of Service _______________   Applicable Skills Learned _______________

What Foreign language do you speak? _______________________   Hobbies or Sports _______________________________

CHARACTER REFERENCES (Give persons who know you well; not previous employers or relatives)

NAME	ADDRESS (Street, City, State)	PHONE NUMBER	OCCUPATION	No. of Yrs. Person has known you

EMPLOYMENT HISTORY
(Start with the most recent employer)

Name and Address of Employer	Employed From To	Kind of Work	Reason for Leaving	Weekly Earnings
Name of Firm: Address Supervisor:	Last Start			Last Start
Name of Firm: Address Supervisor:	Last Start			Last Start
Name of Firm: Address Supervisor:	Last Start			Last Start
Name of Firm: Address Supervisor:	Last Start			Last Start
Name of Firm: Address Supervisor:	Last Start			Last Start

May we contact the employers listed above? _______________ If not, indicate which one(s) you do not wish us to contact _______________________________

____

In case of accident or illness, notify:

Name _______________________________________   Phone _______________________________

Address ____

Expected starting salary _______________________________

What date can you start work? _______________________   If part time, days and hours available _______________________________

How long have you worked with farm equipment? _____________ Do you have a farm background? _____________

Have you worked for other dealers? _____________ What line? _____________

Have you had experience with light industrial equipment?_______ As an operator? _____________

Have you worked for other dealers? _____________ What line? _____________

Have you had experience with lawn and garden equipment?_______ As an operator? _____________

Have you worked for other dealers? _____________ What line? _____________

ACTUAL EXPERIENCE IN ANY OF THE FOLLOWING (Please Check)

SERVICE DEPARTMENT
☐ Service Manager
☐ Shop Foreman
☐ Tractor Mechanic
☐ Implement Mechanic
☐ Hydraulics Mechanic
☐ Small Engine Mechanic
☐ Farmstead Mechanization
   Technician
☐ Machine Set-up
☐ Electrician
☐ Electronics
☐ Diesel Mechanic
☐ Refrigeration
☐ Painting
☐ Truck Driver
☐ Welding

SALES DEPARTMENT
☐ Sales Manager
☐ Farm Machinery Sales
   ☐ New ☐ Used
☐ Light Industrial Equip. Sales
☐ Lawn & Garden Sales
☐ Truck Sales
☐ Auto Sales

OTHER
☐ Janitor
☐ Carpenter
☐ Building Maintenance
☐ Watchman

PARTS DEPARTMENT
☐ Parts Manager
☐ Parts Clerk
☐ Parts Inventory Control

OFFICE
☐ Office Manager
☐ Bookkeeper
☐ Cashier
☐ Secretary-Stenographer
☐ File Clerk
☐ Data Entry Operator

Summarize here any additional experiences and/or skills you may have. _____________________________

____

____

____

The facts set forth in my application for employment are true and complete. I understand that if employed, false statements on this applicaion shall be considered sufficient cause for dismissal. You are hereby authorized to make any investigation of my personal history and financial and credit record through any investigative or credit agencies or bureaus of your choice.

In making this application for employement, I authorize you to make an investigative consumer report whereby information is obtained through personal interviews with my neighbors, friends, or others with whom I am acquainted. This inquiry, if made, may include information as to my character, general reputation, personal characteristics and mode of living. I understand that I have the right to make a written request within a reasonable period of time to receive additional, detailed information about the nature and scope of any such investigative report that is made.

____
Signature of Applicant

This employment application is prepared for general use by equipment dealers and is intended to comply with all federal, state and local fair employment practices laws and with the Fair Credit Reporting Act.

APPLICANT: DO NOT WRITE BELOW THIS LINE
RECORD OF EMPLOYMENT

Date
Started _____/_____/_____   Assigned to: _____________ Starting Salary _____________ Basis of Pay _____________

HISTORY OF JOB ASSIGNMENTS, PAY CHANGES, ETC.

____

____

____

____

____

It pays to select carefully, interview, and hire the best people possible. Carefully cull out unprofitable or questionable people as soon as possible. Be sure your pay policy remunerates your employees adequately.

TRAINING

Training must be a continuing job. In a viable business such as this, you move forward or backward. There is no standing still with yesterday's technology.

Surveys indicate that 65% of lost customers are lost because of employee deficiencies. Training can correct these deficiencies, which will result in increased profits. Training can reduce employee turnover. Many employee shortcomings are really the failure of management to make their needs known.

Training need not scare management. It need not be involved or complicated. It must, however:

Tell the employee
Show the employee
Let him do it
Guide and correct him

This can be accomplished in many ways. Typical of the many methods is the lecture. It is probably not applicable to most local level training. The conference will probably work better. Here a group leader guides the group through a discussion of the subject. There is a feedback between the student and the teacher. Sales training fits this category very well.

The demonstration will be useful in mechanics training. The teacher goes through the assembly and disassembly of a new mechanical unit. It can be combined with the practical exercise. Each student may have a unit to work with. This hands-on training is very good where manual skills are involved. Do not forget on-the-job training. Pair the newcomer with a veteran. When possible, pair him with one who likes to teach, not one who likes to have a fetch-and-carry man around to do his dirty work.

Finally, the employee newspaper or bulletin board can be a valuable training aid. Post pertinent information on the board. New methods of selling or repairing that are novel and work can be passed on very effectively by this method.

Many training aids are available for use on a local level. The first and most important aid is a training plan. Whether formal or informal, decide how the instruction is to be carried out, how much time is to be spent on each phase, and who is to do the instructing. This training plan helps to insure that important parts of the instruction will be carried out.

Individual lesson plans are used in formal instruction, as an aid to the instructor to keep him on course. At the retailer's level, lesson plans will seldom be used. On some subjects, they may be available from manufacturers and should be used when available.

Training films, slides, and charts are usually available from the manufacturer. They give a professional touch to the instruction, which is helpful. Many private groups and governmental agencies also produce films. Sales training is particularly well illustrated in many available films.

For mechanical training, mock-ups and cutaways are very useful. The functioning mechanism can be explained and visualized with these aids.

Observation trips can be quite useful for accounting personnel and for partsmen. It is helpful to visit related facilities where new ideas can be gathered.

Dramatization is an asset to sales training. One can act as the customer, offering the many objections that customers are likely to offer. Various means of overcoming objections, of closing sales, and of opening sales can all be demonstrated. This form of training adds zest to the training.

Do not forget the test as a training aid. It shows the student his weak points. It points out areas where he must do additional work. It also serves as a valuable aid to the instructor to show how effective his instruction has been. It is an aid toward planning further training.

QUESTIONS

1. What is leadership?
2. Define discipline.
3. What is the single most important factor in employee relations?
4. What must a good pay plan do?
5. Compare the profit sharing, bonus, and incentive pay plans.
6. How would you set up a pay plan for your salesmen?
7. What is management?
8. Compare the human approach to the job approach for selecting personnel.
9. What would you look for in a job application form?
10. Compare training salesmen to training mechanics.

14

Merchandising Farm and Power Equipment

The day of casual selling by the amateur is past. Farmers and light industrial contractors are businessmen who expect salesmen to be sales engineers. Sales will be made by men who can justify equipment to the purchaser on the basis of economics. Many sales are still being made on the basis of wants rather than needs, but the complete salesman must be able to operate in both areas. Modern technology demands salespeople who are able to talk in terms of machinery systems for crop and livestock production rather than in terms of single power units.

ORGANIZATION

The full service dealer must departmentalize. The volume of business conducted will determine to some extent the type of organization needed. With few exceptions, three departments—sales, parts, and service—will be needed. The dealer who diversifies will probably need more. A typical organization for a full-service dealer is shown in Fig. 14.1.

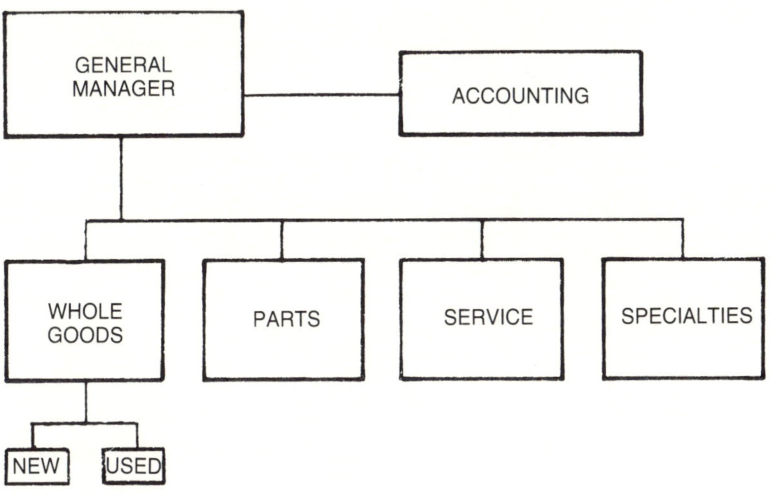

FIG. 14.1. The full-service dealer.

In previous chapters, we have discussed the operation of all departments except whole goods. This is certainly no attempt to minimize its importance. Without a strong whole goods department, it would be very difficult to operate a profitable retail business.

An examination of the 1980 Cost of Doing Business Study (Appendix A) may serve to put its importance into focus. Considering all dealers in North America, new and used sales represented 69.96% of all sales or 42.36% of all margins. It is difficult to assess the net profit this represents. Unfortunately, the expenses in the 1980 study are not broken down by departments. Overall expenses were 14.89% of sales.

The last study that reported expenses by departments was conducted in 1971. The dealers reporting showed margins as 9.43% of sales and whole goods expenses as 11% of sales for a loss of 1.57% of sales. It is evident that the old adage, "All that glitters is not gold," is true in the farm equipment business as well. It would be very shortsighted to consider getting rid of the whole goods department. This is not an acceptable solution. The answer lies in increasing its profitability. Accurate, departmentalized accounting is necessary to do so.

Management

The whole goods department should be headed up by a sales manager. In small organizations, the general manager may well double as sales manager as well. Regardless, there are three main functions that are to be accomplished if profitability is to result. The Sales Manager must:

1. Establish sales goals
2. Direct and assist salesmen toward those goals
3. Establish controls to measure and reward

Goals

Goals may take many forms. The important thing is that they are *measurable* and *attainable*. They should be established based on past performance. The entire sales crew should concur so that they become the goals of the organization rather than the sales manager's goals. To be effective, the goals must be reduced to the contribution of the individual in time units no greater than a month. Goals must be publicized among the participants and monitored so that everyone knows at all times how well he is meeting his goal. A thermometer chart is a simple way of accomplishing this. It is illustrated in Fig. 14.2. Graphs, bar charts, and the like may also be used for this purpose.

Let's look at some goals that fit this category. As illustrated in Fig. 14.2, the goal may simply be a dollar value goal. It is certainly measurable and, if based on past history, it should be attainable. At times of continuing inflation, care must be taken to see that the goal takes inflation into consideration. An 8% increase in sales may be a good goal; but if prices have increased 10%, this goal represents a 2% loss in actual sales. To get an actual increase of 8% during a period of 10% inflation requires an increase of 18.8% over the previous period.

$$\text{GOAL} = [\text{Previous sale} + (\text{Inflation rate} \times \text{Previous sale})] \times (\text{Rate of Increase} + 1)$$
$$\text{GOAL} = [1{,}000{,}000 + 0.10(1{,}000{,}000)]\,(1.08)$$
$$\text{GOAL} = \$1{,}188{,}000$$

The example is based on a past year's sale of $1,000,000, an inflation rate of 10%, and a desired increase of 8%. It is easy to see that this goal lends itself to being monitored monthly and to being subdivided among the individual salesmen.

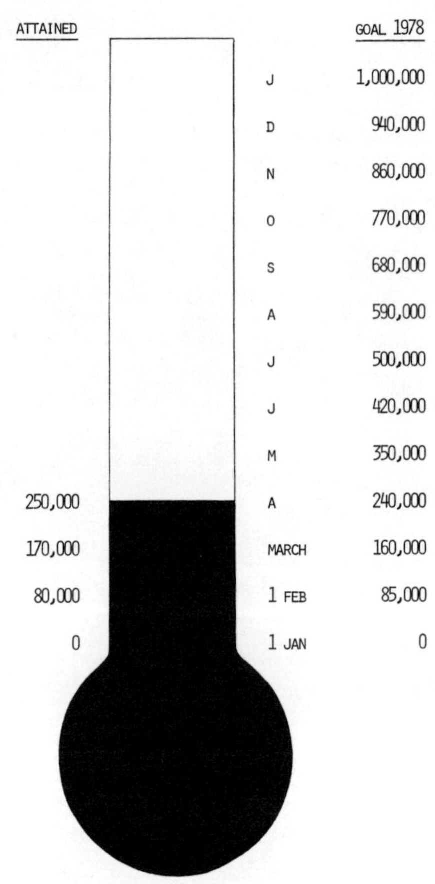

FIG. 14.2. Thermometer-style goal chart, showing monthly goals.

The sales goal may be based on capital invested. Care must be taken to differentiate between equity capital and borrowed capital. Since this varies so widely among dealers, it is harder to generalize. A figure of 5 times invested capital is often used. With a net profit on sales of 4%, this will yield a return on invested capital of 20%. This level should be considered minimal. It can be readily seen that this goal is not as easy to prorate among salesmen as the dollar-volume goal.

Quite often subgoals may be helpful to the sales effort. A goal may be established to penetrate certain geographical areas in order to enlarge the sales territory. Subgoals may be in terms of numbers of certain units that you wish to introduce to an area. As can be seen, goals are versatile and helpful to the sales manager.

DIRECTING AND ASSISTING SALESMEN

The sales manager has the responsibility of assisting and directing his salesmen. There are many tools and techniques available to him for that purpose. Let's consider some of them.

Product literature comes first to mind. It is probably the oldest form of selling aid. Most literature is prepared with the customer in mind. It describes the product. Its content is specifications, capacities, material of design, and other engineering data. The salesman, in addition, needs literature prepared for his use. Because of the complexity of modern farm machinery, additional information is needed on the process for which machines were designed to function. Most manufacturers employ specialists to prepare sales literature for the salesman with that in mind. The salesman is expected to sell value. The salesman must be provided with information to assist him in showing value to the customer.

Additionally, literature is available to the salesman from many public sources dealing with the economic value of machinery. Land grant colleges and the USDA produce countless bulletins, articles, and news releases on new trends in farm mechanization. This literature deals with production practices, economic advantages, and mechanical acceptance of these practices. This literature is not only a valuable training aid for the seller, but is also often an entree to the customer.

Visual aids are devices such as movies, slides, working models, or cut-aways that portray and may explain the salient features of a machine. They can be used to train the salesman and to show to customers. The intricate workings of complicated machines are often very difficult for an individual to explain. Rather than rely on the efforts of a local man, a professional in the film does the explaining. Film has a definite advantage in that mistakes can be edited out. It has the further advantage of being able to show the operation of a machine under varied conditions that may not be locally available.

Movies and slides standardize the story that the manufacturer wants to tell. They tell the basic product story and leave the amplification, local adaptation, and close to the salesman. It reduces selling cost by making selling effort more effective. Unless the product lends itself to this medium, it should not be used. It is not an end in itself, but a means to an end. It can tell the story, but it cannot ask for the order or write it up.

The use of movies and slides in sales training cannot be overemphasized. When these aids cannot be taken to the field, they serve to refresh the memory of the salesman prior to the sales call. Group discussion of the content of the film at the weekly sales meeting is an excellent means of passing on to all the personnel information on how to overcome objections and close more sales.

Sales managers over the years have emphasized the necessity of the demonstration. Taken in its narrowest context it means to show. From this you might picture a machine being hauled to a field to demonstrate how it performs its function. This form of demonstration is expensive and probably of less value today than it has been in the past. Where completely new concepts are involved, it probably is the only way to convince the user. Remember, a poor demonstration is much worse than no demonstration at all.

Organizing a demonstration requires substantial effort. It involves coordination with the service department for transportation of the equipment to be used. An operator must be trained to handle the machinery. Being a good mechanic does not necessarily make you a good machine operator. Quite often, more than one unit of equipment will be required. How to conduct a good demonstration will be covered in Chapter 15.

It is no longer necessary to demonstrate physically every piece of equipment, as it once was, but this type of demonstration still has a place in the business. This is one more area where the sales manager can assist the salesman.

In a broader sense, to demonstrate means to prove beyond doubt. With this definition in mind we can demonstrate in other ways besides physically having a machine operated. We can prove by argument, by example, by experiment, or by logic. Demonstration can be a mental exercise. It has its greatest value as an aid to closing a sale.

The sale of equipment in the 1980s is more closely tied to farm management than to farm production. New systems of farming will involve selling ideas. The proof of the demonstration will have to be the economics of the process. Experiments showing the feasibility of the equipment should be cited. Seldom will it be possible to demonstrate systems physically. Since these systems may involve equipment, field layout, and buildings, the salesman who can plan a smooth transition from the old to the new in steps will be the one who gets the orders signed. The salesman will need help from economists and agricultural engineers.

The sales manager can further assist the salesman with prospect cards. Basically, the salesman should acquire prospects and record them on cards on his own. Management can assist materially by placing sign-in devices near the coffee pot, requiring sign-in cards at all open house events, and having parts and service suggest prospects for various equipment based on sales in their department. These can be processed and put in proper form by the accounting branch under supervision of management. This adds prospects the salesman may have missed. In addition to being sales assistance, the prospect card is also a form of control for management.

ESTABLISHING CONTROLS TO MEASURE AND REWARD

It would be simple if the successful salesman could be measured with one ruler. Unfortunately, this is not the case. The equipment business usually stands on a four-legged stool—new goods, used goods, service, and parts. The mix and quantity of goods sold determine how well the business can be kept on an even keel. The wise manager controls and measures accomplishments from many facets.

The prospect record is basic. The details in the follow-up reports, which are required from the salesman, must be tailored to the individual. Most people detest writing these reports. Many require the discipline these reports demand in order to work all prospects rather than taking the easy way out and working only the more lucrative prospects. Whether or not to require quotas is an individual thing. The important thing is to keep records of calls made and even more important to record the plan for the call-back. These serve as the basis for the weekly review of accomplishments and plans for the coming week.

Profitability of sales can be controlled using several approaches. Since none is infallible, use several of the following. Since net profit is determined by subtracting expenses from margins, this should be the starting point. Direct selling expenses should be maintained at between 4 and 6% of sales. Merchandise should be priced to yield margins of 16−18% of sales. Overall expenses (overhead) must be kept to 13% of sales or less. In order to maintain a decent return on investment, inventory level must be maintained so as to result in a 3 to 5 annual rate of turnover. It is particularly important to control the used inventory. Maximum limits must be set and adhered to. Seldom should used inventory exceed 20% of all inventory. Based on the inventory of new and used equipment alone, used inventory may be as much as 27% of the inventory of new and used equipment. Conditions of limited availability of new equipment may cause these percentages to vary. The manager must recognize real differences in local conditions and adjust for them.

In order to keep inventory levels well balanced, sales must be balanced. Table 14.1 is taken from the 1978, 1979 and 1980 Cost of Doing Business studies. These were three fairly average years, if any years can be considered average.

TABLE 14.1. SALES AND INVENTORIES OF NEW AND USED EQUIPMENT, ALL NORTH AMERICAN OPERATING HIGH-PROFIT DEALERS

Year	Total		New		Used	
	Sales	Inventory	Sales	Inventory	Sales	Inventory
1978	1,443,690	674,705	1,092,772	392,120	350,918	113,955
1979	1,810,372	791,579	1,349,170	497,591	461,202	127,812
1980	1,832,244	970,766	1,404,433	605,737	427,811	144,308
Annual Average	1,695,435	812,340	1,242,125	418,499	413,310	128,691
Turnover		2.32		2.27		2.81

For the 3-year period, for each $4.10 of new sales there was $1.00 of used sales. Used inventory was 25.8% of new inventory or 15.8% of all inventory. Used sales was 24.4% of sales.

Table 14.2 is based on the same 3 years, but is taken from the average of all low-profit dealers.

TABLE 14.2. SALES AND INVENTORIES OF NEW AND USED EQUIPMENT, ALL NORTH AMERICAN OPERATING LOW-PROFIT DEALERS

Year	Total		New		Used	
	Sales	Inventory	Sales	Inventory	Sales	Inventory
1978	1,522,970	586,724	854,316	350,315	293,393	111,543
1979	2,668,056	771,780	1,140,181	472,916	401,595	142,099
1980	1,550,989	783,132	783,608	470,572	283,357	140,047
Annual Average	1,713,991	713,878	923,036	413,267	326,115	131,229
Turnover		2.00		1.98		2.38

For the 3-year period, for each $5.25 of new sales there was $1.00 of used sales. Used inventory was 31.75% of new inventory and 18.4% of all inventory. Used sales was 19.0% of all sales.

The low-profit dealer has too much of his inventory tied up in used equipment, where his margin is barely 4% of sales (see Cost of Doing Business Study). Failure to move used equipment in a timely manner is a big profit-robber. Notice the low turnover of the low-profit dealer.

The sales manager must not only control salesmen but also reward them. A good incentive compensation plan is a sales tool as well as a morale builder. Review compensation plans for the Whole Goods Department found in Chapter 13. Remember, the key to a good plan is one that benefits both the employee and the employer.

SELLING

In a business of this type, selling isn't everything—but nothing happens until something is sold. Too much emphasis cannot be placed on selling. We hear of the born salesman, but most successful salesmen are made, not born. They are made by training. No one can deny that the extrovert may enjoy his job more. In the final analysis, prior preparation and product knowledge contribute more to the successful salesman than the "gift of gab."

Prior Preparation

1. Prior preparation starts with the *prospect cards*. Who needs what? When does he need it? Can we deliver? Well-kept cards that document previous contacts and reactions enables the salesman to direct his sales pitch effectively. There is no groping around for words that will heat up the buying fever.

2. *Plan call backs*. Every sale represents four to five contacts. With every call, plan ahead to the next sales call. This does not mean you should not attempt to close a sale at every visit. It does mean that, if this call fails, you should know and be ready to call back at some future date prior to the customer's buying deadline.

3. *Coordinate your sales calls*. The salesman never has enough time to see all the people who should be seen. Coordinating calls to save time in transit is one of the best ways to stretch selling time. Look at the county map. Plan your calls to minimize travel time.

4. *Rehearse your presentation*. Important sales calls should be rehearsed. Know exactly how you will open the discussion. Plan carefully how you will proceed to convince the customer. Plan at least two ways to close your sale. Try your presentation out on your wife. Rehearsing to yourself as you travel from call to call is always in order. The tape recorder is an excellent means of determining irritating phrases and words that you may have fallen prey to. It is an ideal way to tell if you sound sure of yourself as you make a presentation. There will be many times when the sales pitch will have to be strictly extemporaneous. Prior practice on other calls and items will serve as a good background.

5. *Anticipate his objections*. Every major product is composed of many intricate parts. Some feature of your product will be objected to by some customer. Your company had a reason for choosing that component. If you don't know why, find out. What are some of the typical objections you might run into?

It is too expensive. Find out why yours sells for more than your competitor's, if indeed it does. What superior design features or material make it cost more? The regional service manager, the blockman, and other salesmen with your company can help you answer that question. Remember, this form of objection is not really an objection. The customer wants assurance that he is getting the best price possible and full value for his money. Certainly he is entitled to that.

It is too light. We make it lighter because we use high strength steel. Actually, it is stronger than our competitor's. Or, if the objection applies to a power unit, the answer may be that you have a superior weight transfer mechanism.

The two examples above illustrate the point. Prior to going on your sales call, anticipate the objections that may be raised and, if you don't already know how to answer them, find out beforehand. Remember, if you have to say, "I'll find out and let you know," it means no sale today and another trip back to the farm.

6. *Know the position of the customer.* Know as much about his operation as you possibly can. How else can you discuss with him his size and capacity requirements? Does he have a trade-in? Is it one you can handle? What is his financial position? Will you have to have it financed, or can the customer handle the financing? Will his buying a unit affect the buying habits of other farmers in the area?

7. *Staff anticipated difficulties.* What difficulties might a sale create? There are many. Is someone else promising delivery of the same unit to someone else? When can the unit be assembled and serviced and delivered? Will the finance company add this additional unit to an already extended credit? Will the parts department stock parts for this unit? All these questions and many more affect other departments within the company. Anticipate the difficulties and staff them out before you make promises that cannot be kept. Remember, broken promises don't make friends for the future.

8. *Obtain necessary literature.* Have it with you. It is very convenient to use while explaining features. Leave it with the prospective customer. It will serve to hone his appetite until you return on your next call.

Product Knowledge

It is difficult to overemphasize product knowledge. It must not be limited to the mechanical, the functional, or the engineering. Knowledge must be balanced and complete. Keeping up is almost a full-time job. Let's see what product knowledge must be acquired, and how.

1. *Study the literature.* This material is prepared with much thought and expense. It uncovers the details you miss when you look at an actual machine. Spend time with your competitor's literature. It is a first step in disposing of customers' objections.

2. *Consult your employer or supervisor.* Experience is a wonderful teacher. You learn from it, or you don't last in the selling business. The boss has to know his goods or at least know where to get the information you need.

3. *Consult the blockman or division manager.* Manufacturers' representatives are in contact with the factory, the wholesaler, and other local retailers. The very nature of their job puts them in contact with those who know most about your product line. Their job is to assist you. The good salesman avails himself of this wealth of product knowledge.

4. *The Service Manager.* When it comes to the construction, repair, and maintenance of a machine, the service manager is an ideal source

of information. He is well qualified to advise you about how the machine functions. This is the man who can help you explain why your machine will have less down time, lower maintenance cost, and longer useful life.

5. *Know the specifications*. This can be found in the literature. Fair comparisons cannot be made unless accurate specifications are known.

6. *Know design features*. Engineers spend literally years developing design features. Do not expect to be an amateur designer. Do know the importance of the features. Be able to relate them to greater capacity, lower maintenance, less down time, and so on. Price is nearly always a consideration. Design features are probably the single greatest factor that establishes value.

7. *Know the operation* When you know exactly how your machine works, you are in a position to portray it at its best. Does your plow simply plow, or does it invert the soil incorporating all the organic matter into the soil, or does it lay one furrow slice against the other trapping the organic matter partly in and partly out of the soil? The latter is very important to the moisture-conscious farmer of the West, but might be objectionable to a Mississippi Delta farmer. As simple and as old as the plow is, it is surprising how much can be said about its operation under various conditions. The good salesman learns to relate these operating characteristics to the needs of the community.

8. *Know the maintenance requirements*. How can the machine be kept operating on the farm? Is it greased for life, or is there some fitting that must be lubricated daily? How much down time can be expected for maintenance? Can minor repairs be done on the farm? Machines, of necessity, have become more complicated. Their maintenance need not be.

9. *Know its adjustments*. Sales are often increased when the machine can be adjusted simply. Know these adjustments. A good example of this is the speed adjustment of the combine cylinder. Do you change drive sprockets to change speed, or do you merely move a lever from the driver's seat? Ease of adjustment relates directly to versatility.

10. *Know the benefits*. Very few people trade in for a bigger tractor. They trade for the ability to plow more acres per day, to plow deeper, to pull greater loads, to reduce labor costs, to get more leisure time, to get more timely operations. Learn the benefits to be derived from your machine before you start on that sales call.

11. *Know the cost*. Let's not confuse cost with selling price. Certainly the salesman should know the purchase price of his machine. As important, he should know the cost of operation of the machine.

12. *Know your competition*. Don't knock it—but it won't go away just because you ignore it. You can face up to it if you know its strength and its weaknesses. Literature, technical manuals, and the Nebraska Tractor tests are all solid sources of information available to you. Use them.

13. *Know the tax benefits*. Investment credit laws are constantly being changed according to current economic conditions. Additionally, depreciation techniques vary from time to time. Know these current laws. Combined with a good estimate of a customer's financial position, it can often be shown that the first-year down payment can be completely written off. This is a perfectly legal maneuver that is encouraged by the government, particularly when economic conditions are depressed.

14. *Know the economics*. It has been said that one of the changes to be expected in the 1980s is that farm operators will be more intelligent. It will require salesmen also to be more intelligent. Farmers are concerned with cash flow just as any other businessmen. Sales will be made more on needs than on wants. Prior to making that sales call, take paper and pencil and justify the transaction on an economic basis.

It might appear from the preceding discussion that the salesman might spend hours in preparation prior to hitting the road every morning. It is hoped that preparation once made for a sales call will not have to be prepared totally for every call of a similar nature. Some of the steps will be done in details, but many can be accomplished quickly en route to a call.

Your Approach

The first 10 seconds of a sales call may well be the most important. Certainly, this is the time when the stage is set—for success or for failure. Let us think about the approach.

1. *Know your customer*. Find out all you can about your customer before calling. As a minimum, know his name and how to pronounce it. Your call can be an unwelcome interruption, or it can be opportunity to acquire needed equipment. Make sure your first 10 seconds are at a time that is convenient to the potential customer. You will make mistakes. Be sure you record notes on his card after leaving so that you do not make these mistakes a second time.

2. *Avoid a trite greeting*. "It's sure hot today" is bound to get you a "it sure is," answer, but that does not put you any closer to selling that hay baler.

3. *Get the customer to talking*. "These hot days sure get a fellow to thinking about baling hay," may be a much better way to get the customer talking about your subject for the day, "hay equipment." You can learn most by listening to your customer's views. Only he knows what he wants and what he thinks about hay equipment. Get him to talking about that, and then you can be in a position to proceed with a sales pitch that will sell.

4. *Be pleasant, confident, and congenial*. In the modern lingo, we might say "be cool." Your personal problems have no place at a sales call. Confidence in your ability to present your equipment and your service organization is evident when you are prepared. Be genial, but not to the extent of appearing flighty or silly.

5. *Be neat; be alert*. Dress for the occasion. It's easy to be neat at the office. You perhaps cannot stay as well groomed in the field, but there is no excuse for being sloppy. Be alert so that as the customer talks you can follow his thinking and guide him to those channels you plan to pursue.

Selling Techniques

Volumes have been written on salesmanship. This section is meant to review the more applicable techniques and not to replace a good text on the subject. If you forget all else, remember that the successful salesman is not selling equipment—he sells what equipment can do for the user. Let's see what you should be selling.

Sell ideas about how your product will help buyers do more work—easier, faster, and more economically. You do this by showing the customer the buyer benefits. To do this, you may need to know product features. Let's make this a little clearer by listing some products features and corresponding buyer benefits.

Product feature	Buyer benefit
Weight transfer	Greater no-load fuel economy
	Less wheel slippage
Hydrostatic drive	Stepless speed adjustment
Enclosed cab	Ideal driver environment
Ball bearing	Less down time
Hi tensile steel	Less dead weight

As you can see, each point could be enlarged upon. The product features so proudly pointed out to you by the manufacturer are meaningless to the consumer unless they can be related to buyer benefits. Let's consider some buyer benefits you should be prepared to stress.

Sell savings in maintenance
Sell savings in labor
Sell greater capacity
Sell ease of operation
Sell more leisure time
Sell less repairs
Sell more timeliness of operations
Sell freedom from obsolescence
Sell ease of adjustments
Sell greater precision of operation
Sell pride of ownership
Sell ability to service
Sell better financing
Sell convenience of buying now
Sell related products
Sell better performance

There are many other buyer benefits you can think about. Use any or all of them when they are applicable. As you develop your presentation, *anticipate objections. Invite fair comparisons*, but *get agreements* on your sales points as you go. It is necessary to keep *trading terms* in mind. This means much to the customer and may well be very important to the business as well.

Finally, *ask for the order*. This last step is a most important one. Ask the customer to buy any and every time you feel he has understood your presentation. If he is convinced, you will save much of your time and his. If he is not, you will probably find out what is holding him back. Remember, when you ask for the order, give the customer the opportunity to answer. Various texts list many ways of closing a sale. The following should be helpful:

1. *Committing question method.* Committing questions may take many forms, but they all point in the same direction. You may use one question or several questions. For example, would you prefer the 105 hp model, or do you propose to buy the 120 hp model? Do you want the cab air conditioned? Will you finance with us, or will you provide your own financing? Will delivery tomorrow afternoon around 3:00 p.m. be satisfactory? At times, one question may be in order. Do you want to pick it up at our place, or shall I send it out Monday when our truck is in your area?

2. *Direct Action Method.* This method requires more self-confidence. The salesman must seize an opportunity and take direct action. The conversation may go like this. Customer: "Will you have the new Exacto Planter in stock this spring?" Salesman: "We received a shipment last week. I'll have yours serviced and delivered to you Friday morning."

3. *Lead time closing.* Machinery use is seasonal. In addition, several time frames must usually elapse between the manufacture of a machine and its delivery on the farm. Any of these are suitable for the lead time method. "Shall we start now so that we can get your financing approved?" "Would you like to order now while our shop is not busy so that we can do a better servicing job?"

4. *Special Offer Method.* this method is used when availability or price is in question. It is also used when introducing certain items. It is a sound business transaction when not used as just another excuse to cut prices. Let's illustrate. "To introduce our new foam row marker, we are giving five gallons of concentrate free with each unit sold during the next two weeks."

5. *Pride of Ownership Method.* Each community has one who must have the biggest of the newest unit on the market. Pride of ownership takes precedence over economics. This method stresses the emotional qualities such as prestige of owning and leadership. It is very effective with some people, but may have an adverse effect on others. It should be used with care. The closing statement may go like this. "Buy this land leveler, and you will be remembered around here as the man who brought precision land grading to this community."

6. *Review advantages method.* As you review the advantages of your machine, try to get the customer to agree to their validity. It is often rather difficult for him to back out of buying without contradicting himself.

7. *Forced choice.* Many customers have trouble deciding what to buy. They are indecisive when considering options. Frame the closing question not around a "yes" or "no" answer but around a choice between just two options. For instance, "Would you like us to deliver the model with the Syncromesh Transmission or the Hydrostatic Transmission?"

The preceding techniques must be varied and used with discretion. No one can tell you exactly how to ask Mr. X to buy, but it is a reasonable certainty that Mr. X won't buy until you ask him. The fact that he says no the first time you ask does not mean he won't buy. It merely signals that you have not convinced him yet. Continue your presentation, answering his objections, and ask again for the order. Continue this long enough to get the order if possible but not so long that the customer feels harried. The salesman who can judge that time frame correctly will be the successful salesman.

Follow Up

The post-sale period is an important period. Like the mother who goes into a "post-partum" depression after delivery, many customers suffer similarly after making a purchase. This can be minimized if you thank the customer for his order. Assure him that he has made an excellent buy. Re-sell him when the delivery is made. Be certain that you adhere to the delivery terms you promised when you sold the machine. Follow-through during delivery makes satisfied customers, and satisfied customers are not only repeat customers but the least expensive and most effective advertisement you have in a community. There is no better way to lay groundwork for future purchases.

SELLING USED EQUIPMENT (FIG. 14.3)

Trade-ins must be considered an opportunity, not a problem. Blind acceptance of this axiom, however, could very well cause a problem. It is an opportunity to sell new equipment that would not otherwise be bought as well as an opportunity to sell used equipment. Unless trade-ins are effectively merchandised, the problem of out-of-balance inventory and low cash flow will soon become evident.

Problems with trade-ins have their start when new equipment is sold. *Sell the new equipment first.* Then, *appraise* the trade-in. Don't bid on the trade-in. Apparently, many customers are better salesmen than the salesman himself.

FIG. 14.3. Used equipment can be an opportunity to do business when properly displayed.

Used Equipment Manager

Size of operation and personal preference may dictate whether or not to have a full-time used equipment manager. Regardless of how the department is organized, there are certain duties that have to be performed. Let's look at some of those.

1. *Supervise appraisals.* No job is more important. A competent appraiser is needed if used equipment is to be sold profitably. Costs of sales have to be controlled. The business will be on a sounder basis if only one person does the appraising.

2. *Determine need for reconditioning.* Within the policy established by the dealership, a determination has to be made as to the probable disposal of the trade-in. Can it be reconditioned, guaranteed, or sold as-is—or should it receive minimal repairs and paint and be sold without benefit of guarantee?

3. *Establish selling price.* The manager must set a price that is competitive yet recovers all costs and returns a reasonable profit. Failure to do this results in out-of-balance inventory, overstated inventory, and operational losses.

4. *Display and promote.* There is a constant need for used equipment. The demand varies from nearly new to nearly finished. The manager who displays his array and promotes it will move it.

5. *Control inventory.* Used inventory must turn over at a rate comparable to other inventory. A goal of 3–5 turnovers per year must be maintained.

6. *Control investment.* The manager must balance his capital needs with those of the other departments in the organization.

7. *Coordinate with new sales.* Close coordination will often result in increased new and used sales. The overall good of the organization sometimes dictates whether to push a new sale or to attempt to sell a used item. Inventory levels and investments in used equipment must dictate trade-in policies at any given time. This can be accomplished only with close cooperation.

8. *Sell.* The used equipment manager supervises the sales force. As such, he must establish goals, direct the salesmen and establish controls to measure the efforts and reward them accordingly. This is no different from the sales responsibility of the manager of the whole goods department.

Perhaps we should now look at some of these responsibilities in more detail. The need for reconditioning used equipment must be determined at the time of the trade-in. The work itself must be done promptly. First, until it is reconditioned, it cannot be sold. This means inventory is being carried that is not saleable. In addition, this inventory is getting older, which means it will be worth less when sold. The selling season for used equipment is the same as that for new equipment. Unless it is processed promptly, it may not be available to be sold until the next selling season. The need for quick reconditioning, if that is to be done, cannot be overemphasized. The value of used equipment can be expected to drop 10–15% every year it is not sold. The cost of money tied up will add another 10%, and the storage cost still another 5%. As a minimum, it will cost 25% of the value of a piece of used equipment to carry it over an additional year.

If the decision is to sell as is, clean it up and display it promptly. If an appearance reconditioning only is to be done, paint it, add the minimum work and parts decided upon, and then display it. An appropriate tag should be attached to each piece identifying the conditions under which the equipment is sold.

Conditions of the guarantee should be carefully spelled out. It is important to limit the time as well as the conditions. The guarantee may include all parts and labor. Many dealers find it advantageous to guarantee only a percentage of the parts and labor rather than 100%. When the guarantee is for 100%, some customers may find faults continuously. When they are required to pay 50% of the parts and labor, they may tend to be more conservative.

Establishing the selling price is often more complicated than it sounds. A dealer is entitled to recover all cost and make a reasonable profit. The customer is also entitled to buy where it will be to his best advantage. In pricing used equipment, look around. What are other dealers getting for comparable merchandise? Check the local farm auctions. Look at the official guide published by the National Farm and Power Equipment Dealers Association. The as-is value is the average resale price less average reconditioning cost less 20%. The as-is value corresponds to cost of sale.

Perhaps this can be clarified with an example. The as-is value of a certain tractor is listed as \$596 in the Guide Book. Reconditioning cost is calculated to be \$55. Resale value then is

$$\text{Resale price} = \frac{\text{as-is value}}{1 - \text{margin}\,\%} + \text{Reconditioning cost}$$

$$= \frac{\$596.00}{(1 - 0.20)} + \$55.00 = \$800.00$$

In this example, a margin of 20% is used. This is based on a 15% overhead and 5% for profit. It is not applied to the reconditioning cost, since this is listed at retail.

Profitability of the used department starts with the appraisal. The appraisal should be thorough and complete. Start with an operational test. Use the appraisal form to determine what work is necessary. The flat rate manual is a very good guide to determine labor required to make needed repairs. Parts will have to be estimated. Once the work is done, the exact cost of parts can be quickly determined. The seasonal nature of the product as well as the local demand must be considered. Finally, apply the profit formula:

$$\text{Appraisal} = (\text{Resale price} - \text{reconditioning cost}) - \text{Margin for overhead and profit}$$

Considering all factors, it is believed that a certain used tractor, properly reconditioned, can be sold for \$1600. It will cost \$110 to recondition the unit. What would you appraise it for?

$$\begin{aligned}
\text{Appraisal} &= (1600 - 110) - [(0.20 \times (1600 - 110)] \\
&= 1490 - 298 \\
&= \$1192.00
\end{aligned}$$

Again the illustration is based on 15% overhead and 5% profit. Since most comparisons are made based on sales, there are some who would prefer to make all computations based on total sales. In this case, reconditioning should be based on dealer's cost rather than retail. The appraisal would then be

$$\begin{aligned}
\text{Appraisal} &= 1600 - (0.20 \times 1600) - (88) \\
&= \$1192.00
\end{aligned}$$

The \$88.00 is the \$110.00 reconditioning cost expressed at dealer's cost rather than at retail.

The over-allowance is a potential source of false or bookkeeper's profit. If not properly handled, the value of the inventory is exaggerated, resulting in overstated assets. Probably the simplest way to handle this is to set up an over-allowance as part of the cost of the sale and reduce the inventory by a similar amount. Others choose to list the sale at the amount for which it was made, but list the trade-in at an inventory value as indicated by the Guide. Either method will reflect true conditions in the financial statement. The most important thing is to develop a true value from which to sell used equipment.

Setting the selling price based on a sound appraisal is relatively simple:

$$\text{Selling price} = \text{Appraisal} + \text{Reconditioning} \\ + \text{Overhead} + \text{Profit}$$
$$= (\text{Appraisal} + \text{Reconditioning}) \\ + (0.15 \text{ Selling Price} + 0.05 \text{ Selling Price})$$
$$= 0.20 \text{ Selling Price Price} + \text{Appraisal} \\ + \text{Reconditioning}$$
$$\text{Selling Price} = \frac{\text{Appraisal} + \text{Reconditioning}}{(1 - 0.20)}$$

This is based on 15% overhead and 5% profit. The selling price of a used tractor appraised at $1500 with reconditioning cost of $200 and overhead of 14% of sale and profit of 4% of sale would be:

$$\text{Selling Price} = \frac{1500 + 200}{(1 - .18)} = \frac{1700}{0.82} = \$2073.17$$

It is easier for some people to price merchandise based on cost rather than on selling price. Markup is expressed on cost, and margin is expressed on selling price. To change from one to the other, use the following formula:

$$\text{Markup} = \frac{\text{Margin}}{1 - \text{Margin}}$$

How much would you mark up a used mower if you desire a margin of 20%?

$$\text{Markup} = \frac{0.20}{1 - 0.20} = \frac{0.20}{0.80} = 25 \text{ or } 25\%$$

Knowing the cost (appraisal value plus reconditioning cost), it is quite easy to determine selling price if the desired margin is known.

$$\text{Selling price} = \text{cost} + (\% \text{ markup} \times \text{cost}) \\ = 1700 + 0.2195 \, (1700) \\ = \$2073.17$$

This is the same example as before, using a markup based on 18% margin.

$$\text{Markup} = \frac{0.18}{0.82} = 0.2195 = 21.95\%$$

Once the equipment has been reconditioned and priced, display and promote it. It is well to separate the used equipment from the new. Establish a used lot. It is very desirable to have an all-weather surface. The cost will be repaid in reduced lot maintenance. Equipment sold out of overgrown weeds never brings a fair market price. A lot size of about 250 ft^2 per unit to be displayed should be sufficient. The overall appearance should definitely be that of a sales facility, not a graveyard. Locate it so that it is convenient to a closing facility. A small office is very helpful. Light up the lot not only for security at night but so that customers can see what you have to offer. An elevated place to display your special for the week is always in order.

Used equipment is a necessary adjunct to any equipment dealer. It accounts for from one-fourth to one-third of all equipment sales. Its magnitude alone demands that it be treated as other than a stepchild. Properly handled it will return margins and profits comparable to that of new equipment.

QUESTIONS

1. What are the main functions of the sales manager?
2. Discuss the setting of sales goals.
3. How does the manager assist the salesman?
4. What are the elements of a good demonstration?
5. What controls can be used to assure profitability of sales?
6. Discuss at least three ways of closing a sale.
7. What are the duties of the used equipment manager?
8. How would you set the price on a piece of used equipment?

15

Sales Promotion

Everything you and your employees do or fail to do contributes to the image of your dealership. Image as used here goes far beyond superficial looks. The image you are trying to build is made up in part of the tangible things such as the buildings, grounds, and fleet you operate; but just as important are the intangibles such as the quality of work, employee attitudes, response to emergency calls, empathy, and the feeling of wanting to help that your dealership conveys. Together they make up the mental image that you portray to your customers. Sales promotion starts with creating a "good image." A good service department backed by a well-stocked parts department is the foundation on which to lay this image.

TANGIBLES

A smart-looking sign, visible from a distance, identifies you and your product. It is a constant reminder to the passer-by of your willingness to be of service. The building should be neat and functional. A clean building is the surest indication to customers that you do clean work. A functional building creates the impression of a well-organized and efficient operation.

Your fleet of automobiles, delivery, and repair trucks cover your trade area continuously. Are the drivers courteous? Are the vehicles kept in good repair? Are your company name and product identification prominently displayed on each vehicle? This is good advertisement that you cannot afford to pass up.

The sales lot should not be overlooked (Figs. 15.1 and 15.2). After all, the name of the game is Sell. Separate new from used equipment. Keep junk out of sight. Illuminate the area at night. Change your featured piece regularly. Put it on an all-weather surface. Keep batteries and tires up. Remember the idea is to project an image of an organization ready to go to work for the customer.

FIG. 15.1. A neat display brings in customers.

FIG. 15.2. A covered shed is a great assist to all-weather selling.

INTANGIBLES

The intangibles are just as important but often harder to define. Behave and dress in a socially acceptable manner. This will vary from one area to another. Make sure your organization is a good neighbor. Lend your support to the public-spirited people in your community that work for the good of the community. Crusade, but don't be a crusader.

Develop an "esprit de corps" within your organization. Everyone should be selling the organization at all times. Train your staff to air their clean linen in public and their dirty linen in strict privacy. How can the public believe they are doing business with the right establishment when service blames their failures on parts who in turn blame administration who in turn blame whole goods for having stocked the miserable piece of equipment in the first place? An exaggeration? Unfortunately, it is too often true.

ADVERTISING

The heart of your sales promotion is your advertising campaign. This is the modern way of multiplying your sales effort. Notice that it does not do away with your sales effort. It takes your sales talk in part and places it in thousands of places. There are many good reasons for advertising.

It lets people know that you are in business.
It lets people know what you are selling.
It keeps your firm's name in the customer's mind.
It helps you to create a preference for your brand.
It is an ideal way to make special offers.

Let's see how advertising fits into selling. In order to make a sale we must:

Make contact between the buyer and seller
Arouse interest in your product
Create a preference for your brand
Make a specific proposal
Close the sale

Advertising can do a substantial part of the first three. The last two—the specific proposal and the close—are best done by the salesman.

The Budget

We normally refer to advertising as costing money. If we think positively about this we must instead say that failure to advertise costs money. It is a bona fide cost of doing business, on which we should spend more time trying to increase its usefulness rather than decreasing its cost. Table 15.1 shows the advertising budget by year as a percent of sales of the firms participating in the Cost of Doing Business Study. Over the past 5 years, it can be seen, the average has ranged from a low of 0.53% in 1976 to a high of 0.59% in 1977.

TABLE 15.1. ADVERTISING BUDGET AS A PERCENTAGE OF SALES

Category	1976	1977	1978	1979	1980
All dealers	0.53	0.59	0.53	0.53	0.58
All high profit	0.46	0.50	0.49	0.49	0.53
All low profit	0.61	0.71	0.57	0.60	0.69

There is no positive correlation between profitability and advertising expense. To the contrary, it can be seen that the low-profit dealer spent a greater portion of his budget advertising. Though this data does not show this, low profitability can be attributed to failure to control expenses in general and to maintain margins rather than to excessive advertising. A careful examination of the cost of doing business will show that the dollar value of advertising did not vary greatly between the low-profit and high-profit dealers. From this we might assume that the advertising budget should be about 0.5% of sales. This information cannot be used in this manner. Every business is different. People from different areas respond differently. Base your advertising on your market conditions and business methods. Base your budget on expected sales rather than past sales. This will take care of money needed to increase sales in new areas or new products.

There are three methods generally used to determine the advertising budget.

1. A percentage of sales
2. The objective method
3. A combination of percentage of sales and objective method

The percentage method is clear cut. A sum equal to a specified percentage of sales is agreed upon, and a sales forecast is made. The simplicity of this method makes it very appealing. In the objective method, a target is set and sufficient money is allocated to advertising to accomplish the objective. This is perhaps more realistic, but it is not so easy to follow.

The combination method combines the good points of both previous methods. Based on the conventional business done during the past year, a percentage of sales is allocated to advertising. Any new or additional sales desired are set up as an objective. An advertising campaign is then laid out and costed to accomplish the objective. As an example, let's assume sales during the past year amounted to $1,500,000 and a budget of 0.5% or $7500 is established. In addition, a new line of log skidders is being handled for the first time. There is competition in this field in your trade area. The goal for whole goods in this area for the coming year is set at $250,000. In addition to normal advertising, a demonstration and open house featuring log skidders and other items used in forest mechanization will be held at an estimated cost of $650.00. Adding this $650 to the conventional

0.5% of sales for the projected $250,000 of new forestry mechanization business suggests a budget of $650 plus $1250 or $1900. The total advertising budget for the year would be 0.5% of $1,500,000 plus $1900 or $9400. The flexibility and simplicity of this method is readily apparent.

Do not deceive yourself about advertising. Buying a friend a lunch might get you past IRS, but is it really a bona fide advertising cost? It would be hard to show that this would help accomplish the advertising objective. The following items will more logically increase sales: Display cards — Signs — Posters — Direct Mail — Postage — Mailing List — Handbills — Circulars — Radio — TV — Newspaper — Periodicals — Calendars — Open houses — Movies — Slides and film. Donations, stationary, calling cards, membership in trade associations, and entertaining prospects are costs of doing business but not legitimate advertising expenses.

After deciding how much to spend on advertising, it is necessary to apportion the dollar. Start by setting aside a reserve of about 15% for contingencies. There are many things that can happen during the year that may require a special effort. A flood may alter farming patterns, requiring special effort to push a new tool. Unexpected competition from a new dealer may require special efforts in a trade area. A special effort may be necessary to get a new piece of equipment quickly established in the territory. These are only a few of many things that may require a special effort.

Next divide the remainder into four not necessarily equal parts. Advertising should lead sales. Base the budget on sales volume for each quarter. Spend the most just preceding and during the quarters with the greatest sales volume. Your advertisements might precede peak sales by as much as 60 days. Regardless of when you advertise, charge it to the quarter that you expect to make the sales in.

From the quarterly budget you further subdivide into monthly amounts. Again it is based on expected monthly sales. Study your past sales history. This will show you what items you normally should push during the month. Consider such things as new machines that have just come on the market, and the development of new farming trends. Concentrate your advertisements on big-volume items. This will bring the most customers into the store.

Do not lose track of your emergency fund. If no emergency develops during the first quarter, or the first two quarters, budget a proportionate share to be spent during the subsequent quarters. Remember, the fund is part of your budget and should be spent whether or not an emergency develops.

The equipment dealer sells products and service. Money allotted to each quarter should be subdivided among these. The exact products and services will vary from dealer to dealer. An example listing might look like this:

Tractors	Tillage tools	Drills
Mowers	Balers	Cultivators
Combines	Grain bins	Service
Augers	Miscellaneous equipment	Used goods

With money allocated to products, the budget might look like Table 15.2. Notice that some items are featured year round while others are seasonal. In the service department, advertisements can be used successfully to improve normally slow months. This may be a contra-

TABLE 15.2. ALLOCATING THE BUDGET BY PRODUCTS[1]

Item advertised	Jan. (4%)	Feb. (6%)	Mar. (9%)	Apr. (11%)	May (12%)	Jun. (13%)	Jul. (7%)	Aug. (12%)	Sept. (6%)	Oct. (6%)	Nov. (5%)	Dec. (4%)
Tractors	X	X	X	X	X	X	X	X	X	X	X	X
Tillage tools	X	X	X	X								
Drills		X	X	X	X							
Cultivators			X	X	X	X	X					
Mowers				X	X	X	X	X	X			
Balers					X	X	X	X	X			
Grain bins					X	X	X	X	X	X		
Augers									X	X	X	
Combines						X	X	X	X			
Misc. Equipment	X	X	X	X	X	X	X	X	X	X	X	X
Used goods	X	X	X	X	X	X	X	X	X	X	X	X
Service	X	X	X	X	X	X	X	X	X	X	X	X
Amount per month	$255.00	$382.50	$573.75	$1020.00	$765.00	$828.75	$448.25	$765.00	$382.50	$382.50	$318.75	$255.00

[1]Note: Budget based on 0.5% of a sales volume of $1,500,000, less 15% ($1125.00) for reserves or a budget of $6,375.00.

diction to the previous advice to let the size of the budget float with sales volume. Specials in the service department can and should be used to level off the workload. Use ads to encourage out-of-season repairs.

Now that the budget has been made for the various months, let's turn to the Advertising Work Calendar. One should be made for each month. It isn't necessary to make them up for the year in advance. This calendar will show when, what, where, and how much will be spent on each item. It's an excellent means of keeping within the budget. Using the same example, the calendar for January might look as shown in Table 15.3.

TABLE 15.3. ADVERTISING WORK CALENDAR FOR JANUARY BUDGET OF $255 WITH RESERVE FUND BALANCE OF $1125.00

		Item	How advertised	Planned cost	Actual cost
M	1				
T	2				
W	3	Postage (Tractor flyer)	Direct mail	$ 13.00	$ 15.28
T	4	Tractor cabs and heaters	Newspaper	$ 16.00	$ 16.00
F	5				
S	6				
S	7				
M	8				
T	9	Repair service and parts	Radio	$ 22.00	$ 22.00
W	10	Repair service and parts	Classified	$ 6.00	$ 6.00
T	11	Rotary choppers	Newspaper	$ 16.00	$ 16.00
F	12	Repair service and parts	Radio	$ 22.00	$ 22.00
S	13				
S	14				
M	15				
T	16	Repair service and parts	Radio	$ 22.00	$ 22.00
W	17	Repair service and parts	Newspaper	$ 16.00	$ 16.00
T	18	Disc Plows	Classified	$ 6.00	$ 6.00
F	19	Repair service and parts	Radio	$ 22.00	$ 22.00
S	20				
S	21				
M	22				
T	23	Repair service and parts	Radio	$ 22.00	$ 22.00
W	24	Repair service and parts	Classified	$ 6.00	$ 6.00
T	25	Tractors	Newspaper	$ 16.00	$ 16.00
F	26	Repair service and parts	Radio	$ 22.00	$ 22.00
S	27				
S	28				
M	29				
T	30	Used Disc Harrows	Radio	$ 22.00	$ 22.00
W	31	Repair service and parts	Classified	$ 6.00	$ 6.00
			Total	$255.00	$257.28

The Media

There is a wide array of advertising media available to the businessman. Select the ones that will best reach your customers. The media must (a) reach the market, (b) cover it thoroughly, and (c) work it regularly. In addition the service must be affordable. All things considered, newspapers, radio, television, and direct mail probably meet the criteria best.

Newspapers.   Every trade area has its favorite newspaper. Selecting a paper becomes a problem in areas less well defined where two or more papers cover the area. In placing ads in the paper, try always to get the same page. If the paper features an agricultural section, place your ad there. Make your layout distinctive. A logo for your business could be quite helpful. Use the classified section for used equipment. It is economical and widely read by customers.

Radio and Television.   Radio is a natural. Tractors equipped with cabs usually have a radio. What an opportunity for a captive audience. During the busy season, 5:30 AM and 12 to 1 are popular radio listening times.

When possible, tie your commercials to existing farm programs. Farmers enjoy news, market and weather reports, and other information programs dealing with agriculture. In the entertainment area you will find farmers partial to folk, country and western, and religious music.

Television is a powerful medium. The service is not inexpensive, however. Investigate possibilities in your area. A tie-in with major manufacturers may be possible. This medium has replaced the radio for home listening in many areas.

Direct Mail.   Direct mail in advance of the main selling season can be very effective. Manufacturers furnish sales literature for this purpose. Enclose a piece with every statement sent out. This gives it a free ride. Mail descriptive literature to a customer in advance of a sales call. This gets the customer thinking about the equipment and makes selling easier. Your receivables listing makes a good starting mailing list. In a new area where you have no list, you can send advertising matter to farmers with nothing more than the RFD number on the envelope. Talk to your postmaster about details and instructions.

The mainstay of most farm equipment dealers' advertising is made up of newspaper, radio, and direct mail advertising. Other means should not be overlooked. Effective signs on the store and on trucks must be considered. A display ad in the phone book is a good investment. Calendars, book matches, yardsticks, and the like are all means of keeping your company name in front of the customer. Avoid scattering your effort too widely. Concentrate on those efforts that pay off most in your area.

If you can sell, you should be able to write advertising copy. Basically, the techniques are very similar. You must make a contact between the buyer and the seller. You must create a preference for a brand, and you must arouse the buyer's interest in the product. What better way than stressing user benefits, which is the basis of sales. Your customers are businessmen who wish to save time and labor and to do jobs more economically. The equipment dealer has a big advantage, since the equipment he sells is self-liquidating. Your ads should constantly stress the high living standard farmers enjoy because of their use of modern equipment.

The Meritt Owens Advertising Agency has worked out a list of 15 signposts to guide advertising copy writers. It can be used as a checklist.

1.   Statement of news value and product in headline
2.   Curiosity-provoking headline
3.   Self-interest appeal in headline and illustration
4.   Testimonial or case history in headline and copy
5.   Copy appeal aimed specifically to prospect with "You" attitude throughout
6.   Attention-getting layout with dramatic illustration
7.   Product illustrated in use
8.   Carry-through of headline, illustration, and lead
9.   Adequate copy to tell story
10.   Direct urge for action
11.   Inclusion of signature or logo-type
12.   Reasons for use of product, in copy
13.   Mention of cost or price, in copy
14.   Mention of product three times—head, lead, copy
15.   How to obtain product or service

Remember, it is not only necessary to tell the story but also necessary to repeat it week after week. There are many good books on the subject. With their guidance you too can be effective. Remember it is basically a sales presentation—something you do every day.

Displays

Displays are a basic building block in your sales promotion campaign. Their main purpose is to create a desire and cause the customer to fulfill it. The display is good advertising because it focuses on customers on your premises. It creates the impulse to buy where the customer is in the right place to buy. According to *Farm and Power* in their "Guide to Profitable Displays" you must display it if you want to sell it.

There are several basic rules that apply. Keep these in mind:

1.   Feature equipment sold.
2.   Put human interest in the display.
3.   Group related items.
4.   Change frequently.
5.   Keep it simple.
6.   Keep it clean and well-lighted.
7.   Plan ahead.
8.   Tie it in with national advertising.

Whether the display is indoors or outside, the basic rules apply. Unless your facilities encourage the customer to browse through the display it will serve no useful purpose. The all-weather surface, the environmental comfort, the coffee pot, and the restroom must all be considered when capitalizing on display selling.

Open House

The open house has been popularized during the past few years. Properly planned, it is an excellent means of acquainting your customers with your products. As it is not inexpensive, you may want to tie in your suppliers with the open house. Unless you have a reason for inviting people, don't. The open house can be successful only if you have a carefully planned program designed to satisfy an objective.

What is the objective of the open house? Is it to introduce new models, to introduce a new line, to show new facilities, to dedicate a new building, or to thank old friends for past business? Whatever it is, plan a program to suit the occasion. You do not invite people in just to see a business-as-usual operation. After all, they can do that any day of the week. You should have an objective and a plan to reach that objective.

The open-house date is most important. Do not take it lightly. Many people are involved. On what day can they all attend? Check with speakers, with suppliers, with the Chamber of Commerce, and with local agricultural agencies to make sure that conflicts can be kept to a minimum. Once the date is set, announce it early so that others are discouraged from scheduling events on that date. The crowd you get for your first one will depend on the advertising you do. The crowd you get for subsequent ones will depend on past programs you have held and the advertising you do. Competition for peoples' time is keen. Do not be guilty of wasting their time.

Plan a program of interest to people. Arrange your facilities to make people comfortable. Make it convenient for them to see and hear the things you want them to see. Use this day as a good way of improving your mailing list. Have everyone register. Offering door prizes is a way to encourage registration. The registration card becomes a prospect card, if you remember to ask on it what particular implement they plan to buy next. Greet the people as they come in. After all, you invited them. Be a good host.

The cost of the open house can vary considerably according to its objective. It may be a simple affair with cold drinks and hot dogs or a more elaborate affair with sit-down dining. Whether it is simple or more elaborate, it is a good sales promotion affair when well planned.

Demonstrations

The demonstration is another technique of sales promotion. It is not as useful today as it was when mechanization was just being introduced to American farmers. When new or revolutionary concepts are being introduced it is a most effective method of selling. To routinely demonstrate equipment found in a trade areas is too expensive to be practical. When considering a demonstration, bear in mind this old axiom, "If anything can go wrong, it will."

There are six elements of a good demonstration.

1.   Know the answers. It is necessary to prepare yourself thoroughly for the demonstration. Know everything there is to know about the equipment. Know its features, its economics, its value to the customer. Know your competitors, too.

2.   Plan the demonstration. Decide how to show the equipment at the appointed place. Plan how you will run the machine through its paces, how long you will operate and how you will show its superiority over its competitor.

3.   Practice. A plow has to be set. Soil types vary. The growing habits of crops are different from year to year. A good shop mechanic

may well be a poor equipment operator. If the demonstration is worthwhile doing, then it is worthwhile practicing before an audience arrives.

4. Let him do it. To the extent that is possible, let the customer operate the machine. Pride of ownership and accomplishment plays a vital part.

5. Make sure he understands. Give the customer an opportunity to ask questions. Find out if he understands. Verify that your sales points are well understood by the customer before leaving.

6. Clean up your mess. Do not leave the premises in an untidy condition. Finish the cut or square or at least stop at an acceptable point. Your plan should have considered what to do with the by-products of the demonstration if there are any.

CONCLUSION

Sales promotion is an ever−present task. It is a pleasant task. It is doing the many things it takes to do what you do best-selling. There is a tendency in the retail equipment industry to use the "shotgun" approach in advertising. Ads are directed at a diverse audience and cover a wide range of products. There is some merit in this. Consider, however, making greater use of the rifle approach. Identify your target, then aim your advertising campaign directly at the target. A good hunter knows that the shotgun is great for small birds, but the big game is bagged only with the rifle.

QUESTIONS

1. What intangible things go to make up the image of a dealership?
2. What is necessary to make a sale?
3. What part of the sales effort can advertising do?
4. What methods can be used to set the advertising budget?
5. Discuss how to make an advertising budget.
6. What are the merits of an open house?
7. How can you make displays more effective?

1980 Cost of Doing Business Study

The National Farm and Power Equipment Dealers Association conducts an annual study of the significant parameters affecting the operations of farm and power equipment retailers. Approximately 1450 U.S. and Canadian dealers participate. Dealers are divided according to profitability. The high-profit dealers are composed of the upper 25% of the sample. The middle 50% are the average dealers and the lowest 25% are the low-profit dealers.

All percentages shown are in relation to sales. Operating percentages are important since they reflect the degree of operating efficiency. A suitable breakdown is used so that departmental figures can be evaluated. Personnel studies are included for guidance in staffing a dealership.

The 1980 study is reproduced in this Appendix as resource material for students.

DEALER SALES TRENDS

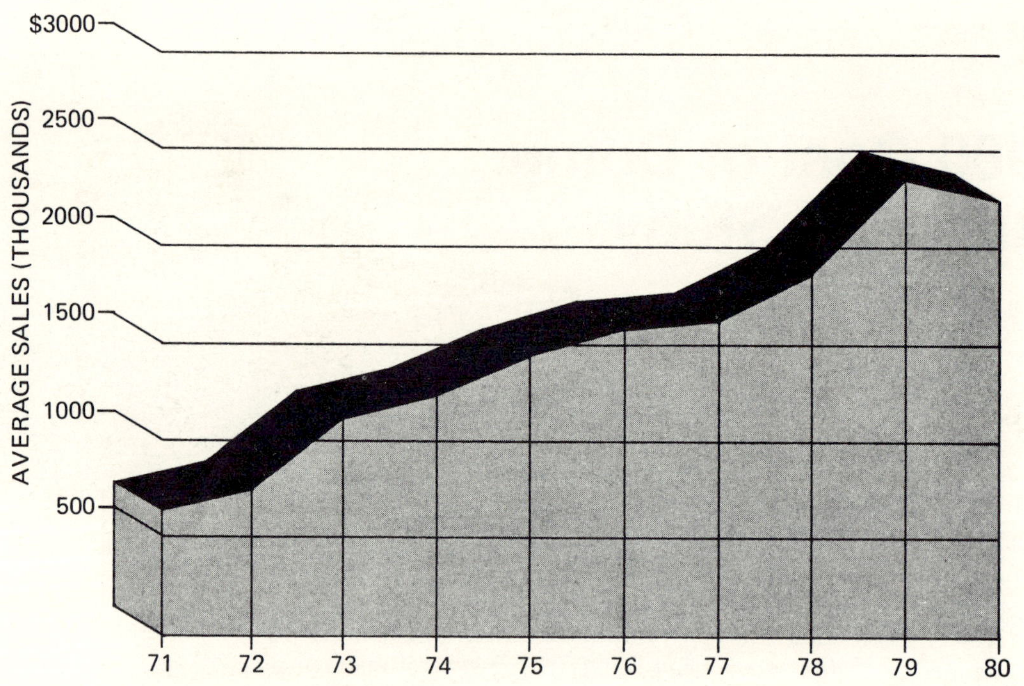

DEALER MARGIN AND PROFIT TRENDS

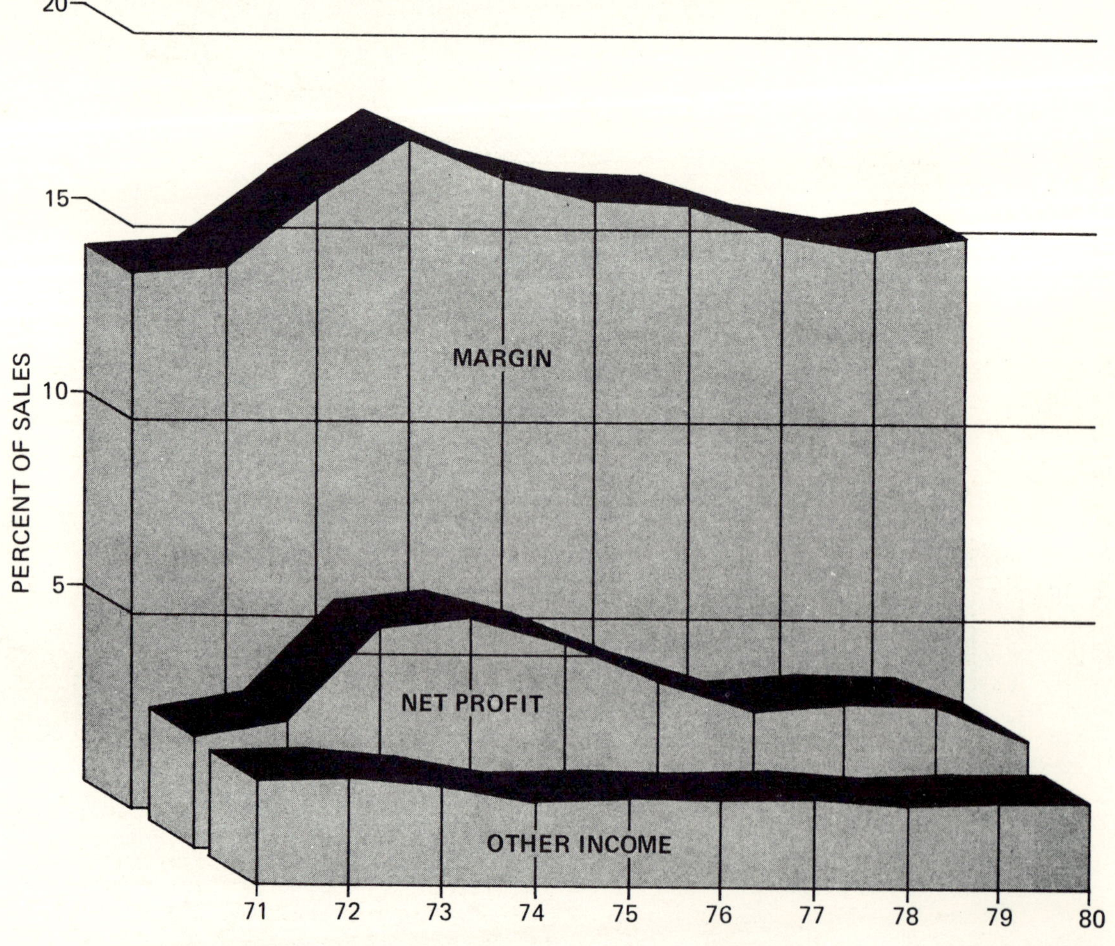

WHOLE GOODS SALES TRENDS

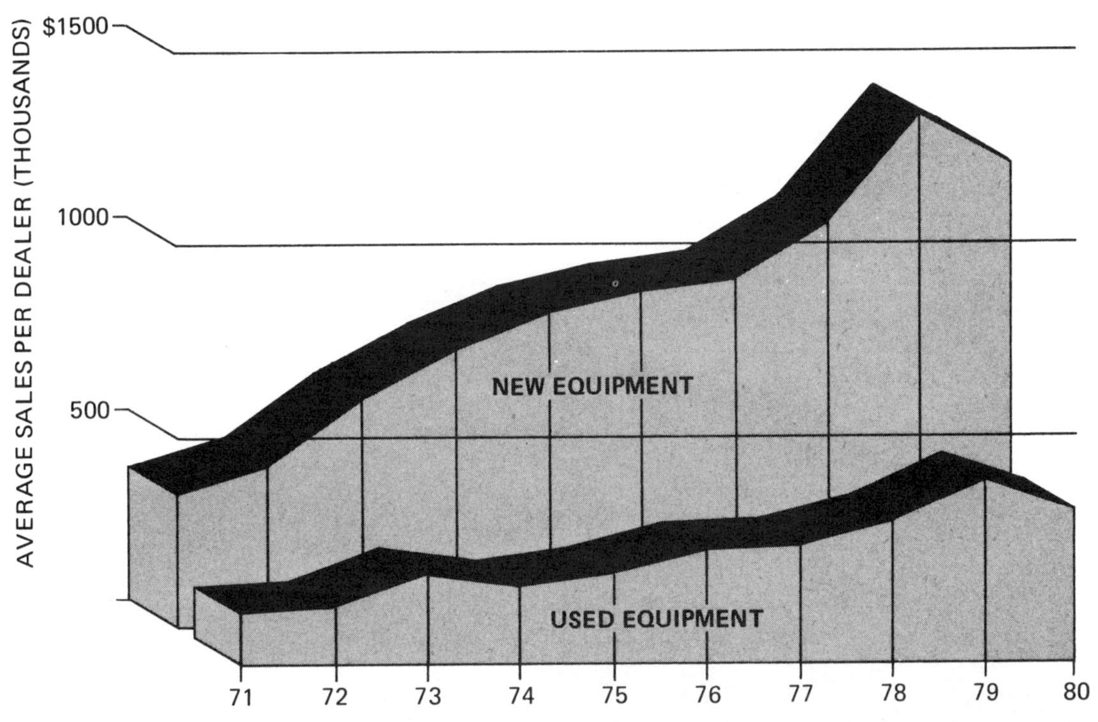

PARTS-SERVICE-OTHER SALES TRENDS

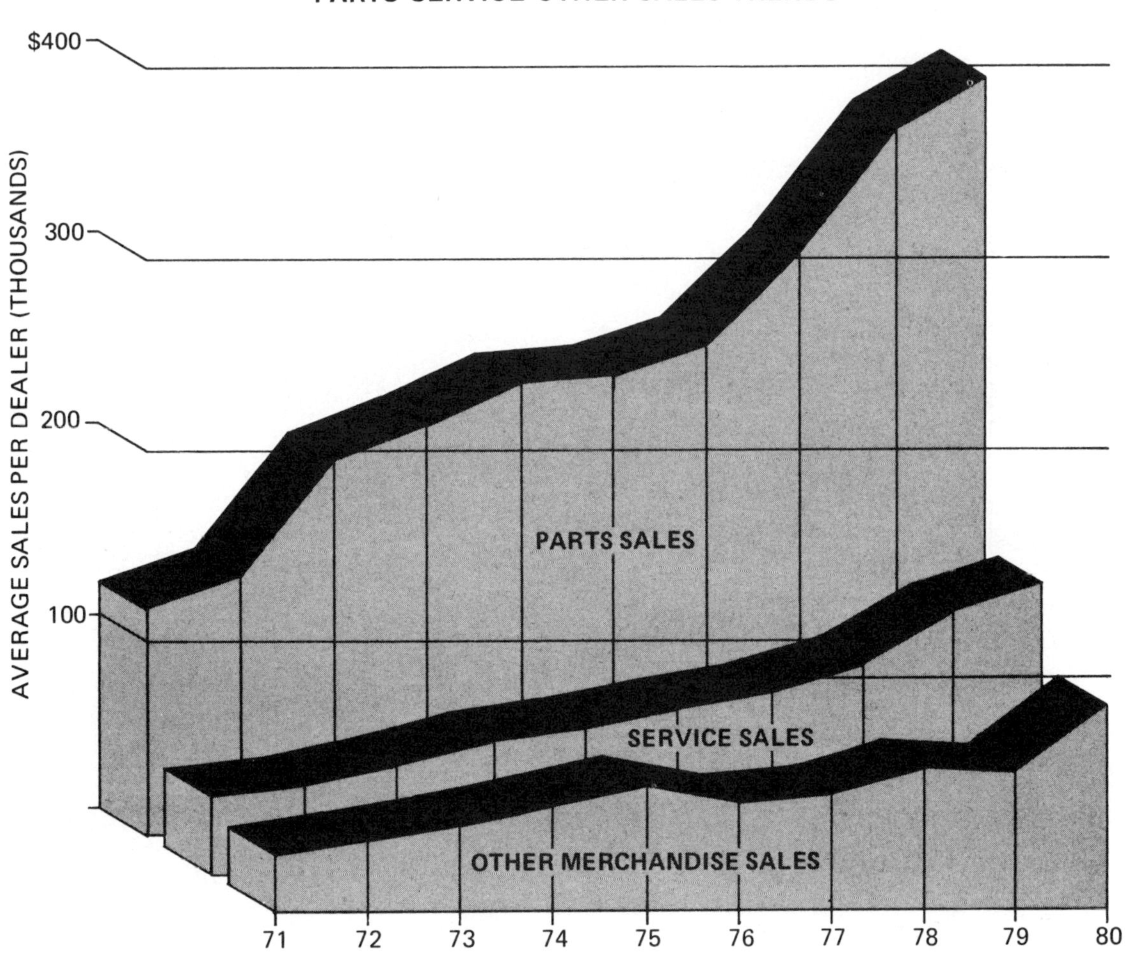

DEPARTMENTAL MARGIN TRENDS

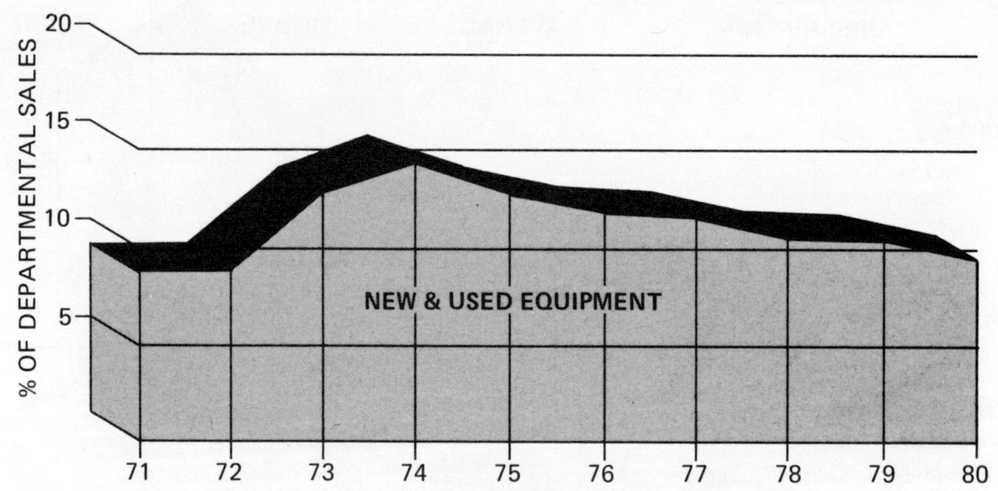

DEPARTMENTAL MARGIN TRENDS

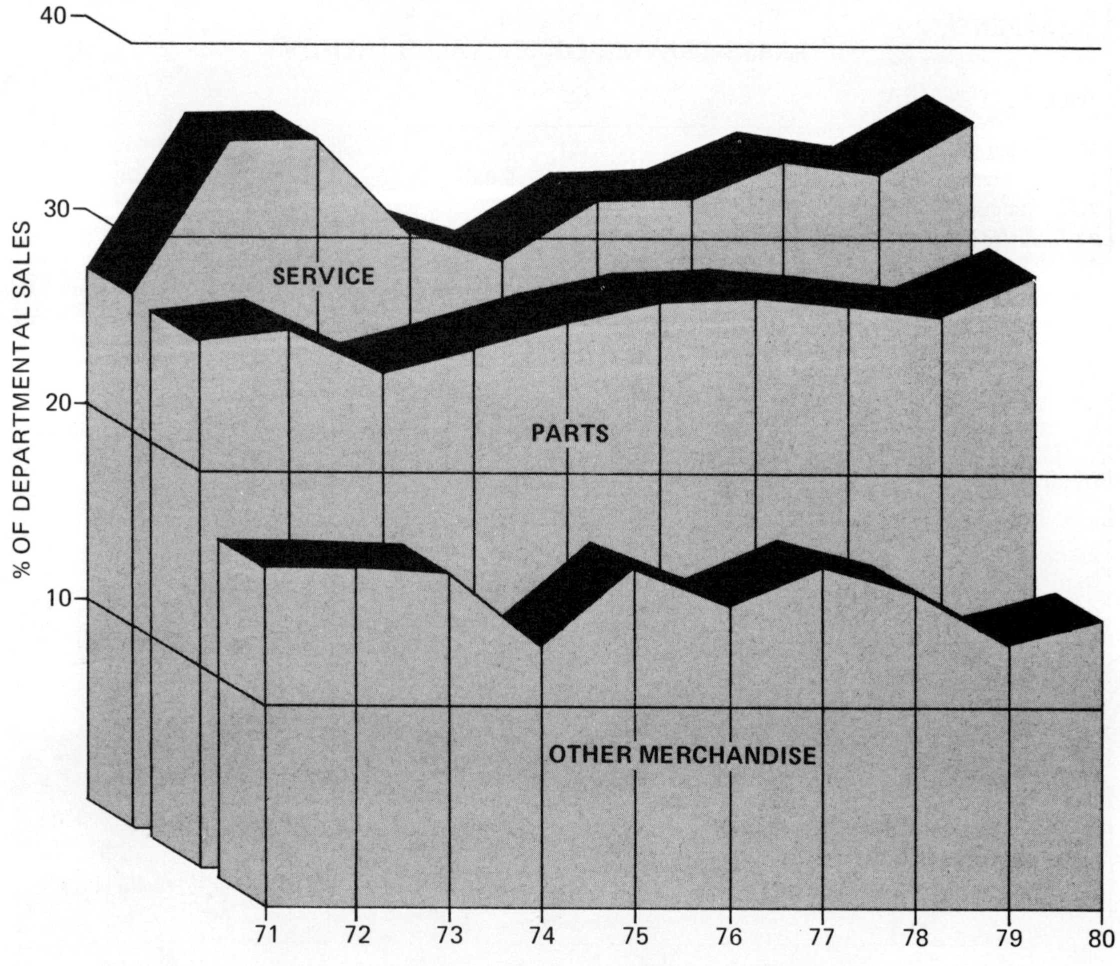

NORTH AMERICAN OPERATING AVERAGES
ALL DEALERS

LINE NO.	ACCOUNT DESCRIPTION	AVERAGE		YOUR FIGURES		HIGH PROFIT		LOW PROFIT	
		Amount	% Sales	Amount	% Sales	Amount	% Sales	Amount	% Sales
	SALES:								
1	New Equipment	$1,176,693	52.24			$1,404,433	52.47	$ 783.608	50.52
2	Used Equipment	399,201	17.72			427,811	15.98	283,357	18.27
3	*Total New and Used*	1,575,894	69.96			1,832,244	68.45	1,066,965	68.79
4	Repair Parts	395,027	17.54			505,190	18.87	276,027	17.79
5	Service Labor	149,875	6.66			184,786	6.91	109,249	7.05
6	All Other Lines	105,248	4.67			127,533	4.76	77,059	4.97
7	Rental-Lease Income	26,262	1.17			26,824	1.01	21,689	1.40
8	**TOTAL SALES**	2,252,306	100.00			2,676,577	100.00	1,550,989	100.00
	MARGINS:								
9	New Equipment	118,636	10.08			170,139	12.11	67,065	8.56
10	Used Equipment	28,163	7.05			49,335	11.53	13,796	4.87
11	*Total New and Used*	146,799	9.32			219,474	11.98	80,861	7.58
12	Repair Parts	121,658	30.80			160,391	31.75	82,660	29.95
13	Service Labor	54,681	36.48			71,979	38.95	32,749	29.98
14	All Other Lines	15,737	14.95			22,474	17.62	10,225	13.27
15	Rental-Lease Income	7,666	29.19			8,791	32.77	5,492	25.32
16	**TOTAL OPERATING MARGIN**	346,541	15.39			483,109	18.05	211,987	13.67
	EXPENSES:								
17	Salaries-Salesmen	30,816	1.36			37,223	1.39	21,596	1.39
18	Travel and Training	4,675	.21			5,754	.21	3,683	.23
19	Advertising and Promotion	13,065	.58			14,144	.53	10,586	.69
20	After Sales & Warranty Exp.	15,381	.68			18,499	.69	11,648	.75
21	**TOTAL SALES EXPENSES**	63,937	2.83			75,620	2.82	47,513	3.06
22	Salaries - Officers, Owners	40,368	1.80			44,994	1.68	27,783	1.79
23	Salaries - Partsmen	28,683	1.27			35,229	1.32	21,202	1.37
24	Salaries - Office and Others	24,114	1.07			27,756	1.03	18,572	1.19
25	Occupancy Expense	32,855	1.46			34,518	1.29	27,756	1.79
26	Heat, Light, Power, Water	6,684	.30			6,829	.26	5,984	.39
27	Telephone and Telegraph	5,469	.24			6,206	.23	4,632	.30
28	Office & Store Exp., and Postage	10,820	.48			11,996	.45	8,074	.52
29	Shop Supplies and Expense	9,322	.41			10,963	.41	7,383	.47
30	Car and Truck Expense	21,362	.95			25,297	.94	17,555	1.14
31	Taxes - Except Payroll & Real Estate	4,481	.20			5,168	.20	3,472	.22
32	Employee Benefits	28,453	1.26			33,885	1.26	20,704	1.34
33	Dues and Subscriptions	830	.04			914	.04	732	.04
34	Insurance - Except Group & R.E.	11,492	.51			12,808	.48	10,293	.67
35	Legal and Auditing	3,291	.15			3,753	.14	3,168	.20
36	Interest & Bank Chgs - Except R.E.	24,385	1.08			17,668	.66	30,929	1.99
37	Depreciation — Except R.E.	11,900	.53			12,314	.46	9,472	.62
38	Bad Debts	3,313	.14			3,553	.13	1,882	.12
39	Contributions and Miscellaneous	3,807	.17			3,835	.14	2,503	.16
40	**TOTAL EXPENSES**	335,566	14.89			373,306	13.94	269,609	17.38
41	**NET PROFIT ON SALES**	10,975	.49			109,803	4.10	(57,622)	(3.71)
42	**Other Income**	53,746	2.38			73,464	2.74	28,815	1.85
43	**NET OPERATING PROFIT OR LOSS (Before Income Taxes)**	64,721	2.87			183,267	6.84	(28,807)	(1.86)

REGIONAL OPERATING AVERAGES
ALL DEALERS

LINE NO.	ACCOUNT DESCRIPTION	EASTERN		YOUR FIGURES		CENTRAL		WESTERN	
		Amount	% Sales	Amount	% Sales	Amount	% Sales	Amount	% Sales
	SALES:								
1	New Equipment	$ 909,794	47.94			$1,142,997	52.85	$1,417,009	53.51
2	Used Equipment	271,231	14.30			422,810	19.55	447,416	16.90
3	*Total New and Used*	1,181,025	62.24			1,565,807	72.40	1,864,425	70.41
4	Repair Parts	387,160	20.40			352,412	16.30	472,249	17.84
5	Service Labor	144,862	7.64			142,835	6.61	165,187	6.23
6	All Other Lines	150,614	7.94			83,223	3.84	111,169	4.20
7	Rental-Lease Income	33,760	1.78			18,177	.85	34,729	1.32
8	**TOTAL SALES**	1,897,421	100.00			2,162,454	100.00	2,647,759	100.00
	MARGINS:								
9	New Equipment	101,085	11.11			110,042	9.63	145,185	10.25
10	Used Equipment	12,382	4.57			27,081	6.41	40,840	9.13
11	*Total New and Used*	113,467	9.61			137,123	8.76	186,025	9.98
12	Repair Parts	117,503	30.35			108,432	30.77	146,803	31.09
13	Service Labor	51,971	35.88			51,825	36.28	61,357	37.14
14	All Other Lines	23,583	15.66			12,369	14.86	16,020	14.41
15	Rental-Lease Income	7,678	22.74			7,271	40.00	8,322	23.96
16	**TOTAL OPERATING MARGIN**	314,202	16.56			317,020	14.66	418,527	15.81
	EXPENSES:								
17	Salaries-Salesmen	29,162	1.53			26,975	1.24	38,425	1.45
18	Travel and Training	4,248	.23			4,064	.19	6,001	.22
19	Advertising and Promotion	11,321	.59			12,683	.59	14,907	.57
20	After Sales & Warranty Exp.	10,889	.58			14,210	.65	20,445	.77
21	**TOTAL SALES EXPENSES**	55,620	2.93			57,932	2.67	79,778	3.01
22	Salaries - Officers, Owners	38,349	2.02			38,302	1.78	45,238	1.71
23	Salaries - Partsmen	26,516	1.39			26,463	1.22	33,912	1.28
24	Salaries - Office and Others	25,914	1.37			21,154	.98	27,865	1.05
25	Occupancy Expense	25,645	1.35			32,166	1.48	38,973	1.47
26	Heat, Light, Power, Water	6,440	.34			6,832	.32	6,604	.25
27	Telephone and Telegraph	5,653	.30			5,034	.23	6,076	.23
28	Office & Store Exp., and Postage	11,240	.59			9,657	.45	12,491	.47
29	Shop Supplies and Expense	8,404	.44			8,505	.39	11,329	.43
30	Car and Truck Expense	21,211	1.12			20,053	.93	23,673	.89
31	Taxes - Except Payroll & Real Estate	5,167	.28			4,063	.19	4,715	.18
32	Employee Benefits	27,428	1.44			26,899	1.24	31,777	1.20
33	Dues and Subscriptions	828	.04			784	.04	910	.04
34	Insurance - Except Group & R.E.	10,727	.57			11,445	.53	12,097	.45
35	Legal and Auditing	3,299	.17			3,043	.14	3,704	.14
36	Interest & Bank Chgs - Except R.E.	25,211	1.33			24,746	1.14	23,209	.88
37	Depreciation — Except R.E.	11,099	.59			11,041	.51	13,897	.53
38	Bad Debts	4,959	.26			2,465	.12	3,610	.13
39	Contributions and Miscellaneous	3,254	.17			3,894	.18	4,041	.15
40	**TOTAL EXPENSES**	316,964	16.70			314,478	14.54	383,899	14.49
41	**NET PROFIT ON SALES**	(2,762)	(.14)			2,542	.12	34,628	1.31
42	**Other Income**	50,575	2.66			47,999	2.21	65,612	2.47
43	**NET OPERATING PROFIT OR LOSS (Before Income Taxes)**	47,813	2.52			50,541	2.33	100,240	3.78

GROUP "A" OPERATING AVERAGES
DEALERS WITH SALES TO $1,000,000

LINE NO.	ACCOUNT DESCRIPTION	AVERAGE Amount	AVERAGE % Sales	YOUR FIGURES Amount	YOUR FIGURES % Sales	HIGH PROFIT Amount	HIGH PROFIT % Sales	LOW PROFIT Amount	LOW PROFIT % Sales
	SALES:								
1	New Equipment	$367,096	50.93			$437,944	51.62	$296,226	48.78
2	Used Equipment	106,760	14.81			126,982	14.97	96,244	15.85
3	*Total New and Used*	473,856	65.74			564,926	66.59	392,450	64.63
4	Repair Parts	160,671	22.29			176,460	20.80	146,071	24.05
5	Service Labor	49,245	6.83			58,959	6.95	43,586	7.18
6	All Other Lines	28,349	3.93			35,451	4.18	22,034	3.63
7	Rental-Lease Income	8,657	1.21			12,518	1.48	3,064	.51
8	**TOTAL SALES**	720,778	100.00			848,314	100.00	607,205	100.00
	MARGINS:								
9	New Equipment	41,819	11.39			58,003	13.24	26,611	8.98
10	Used Equipment	9,087	8.51			13,023	10.26	8,635	8.97
11	*Total New and Used*	50,906	10.74			71,026	12.57	35,246	8.98
12	Repair Parts	50,730	31.57			57,856	32.79	45,957	31.46
13	Service Labor	11,018	22.37			20,013	33.94	1,057	2.43
14	All Other Lines	4,451	15.70			7,708	21.74	3,164	14.36
15	Rental-Lease Income	1,920	22.18			2,314	18.49	2,332	76.11
16	**TOTAL OPERATING MARGIN**	119,025	16.51			158,917	18.73	87,756	14.45
	EXPENSES:								
17	Salaries-Salesmen	8,611	1.19			9,749	1.14	7,981	1.31
18	Travel and Training	1,664	.23			1,540	.19	1,847	.30
19	Advertising and Promotion	6,037	.84			6,732	.79	5,943	.98
20	After Sales & Warranty Exp.	3,048	.42			2,767	.33	3,029	.51
21	**TOTAL SALES EXPENSES**	19,360	2.68			20,788	2.45	18,800	3.10
22	Salaries - Officers, Owners	15,623	2.17			17,563	2.07	13,808	2.27
23	Salaries - Partsmen	10,760	1.49			10,977	1.29	10,630	1.75
24	Salaries - Office and Others	8,608	1.20			9,758	1.15	8,743	1.44
25	Occupancy Expense	13,262	1.84			15,242	1.80	14,185	2.33
26	Heat, Light, Power, Water	3,550	.49			3,288	.38	3,967	.66
27	Telephone and Telegraph	2,324	.32			2,320	.28	2,543	.41
28	Office & Store Exp., and Postage	3,361	.47			2,946	.35	3,951	.65
29	Shop Supplies and Expense	3,852	.53			4,182	.49	4,468	.74
30	Car and Truck Expense	8,370	1.16			8,800	1.04	8,146	1.34
31	Taxes - Except Payroll & Real Estate	2,131	.30			2,634	.31	2,028	.34
32	Employee Benefits	8,865	1.23			8,581	1.01	8,644	1.42
33	Dues and Subscriptions	447	.06			430	.05	467	.08
34	Insurance - Except Group & R.E.	5,230	.73			4,403	.52	5,556	.91
35	Legal and Auditing	1,756	.24			1,702	.20	1,738	.29
36	Interest & Bank Chgs - Except R.E.	10,537	1.46			5,176	.61	15,576	2.56
37	Depreciation — Except R.E.	4,537	.63			4,330	.51	4,985	.82
38	Bad Debts	915	.13			1,298	.15	651	.11
39	Contributions and Miscellaneous	1,283	.18			1,199	.14	1,499	.25
40	**TOTAL EXPENSES**	124,771	17.31			125,617	14.80	130,385	21.47
41	**NET PROFIT ON SALES**	(5,746)	(.79)			33,300	3.93	(42,629)	(7.02)
42	Other Income	16,576	2.29			23,759	2.80	11,303	1.86-
43	**NET OPERATING PROFIT OR LOSS (Before Income Taxes)**	10,830	1.50			57,059	6.73	(31,326)	(5.16)

GROUP "B" OPERATING AVERAGES
DEALERS WITH SALES $1,000,000 TO $2,000,000

LINE NO.	ACCOUNT DESCRIPTION	AVERAGE		YOUR FIGURES		HIGH PROFIT		LOW PROFIT	
		Amount	% Sales	Amount	% Sales	Amount	% Sales	Amount	% Sales
	SALES:								
1	New Equipment	$ 769,709	52.60			$ 784,541	54.41	$ 714,424	49.75
2	Used Equipment	249,082	17.02			218,487	15.16	265,466	18.49
3	*Total New and Used*	1,018,791	69.62			1,003,028	69.57	979,890	68.24
4	Repair Parts	278,700	19.04			294,381	20.41	259,095	18.05
5	Service Labor	94,868	6.49			85,276	5.92	106,358	7.41
6	All Other Lines	59,279	4.05			47,406	3.29	74,779	5.20
7	Rental-Lease Income	11,668	.80			11,646	.81	15,657	1.10
8	**TOTAL SALES**	**1,463,306**	**100.00**			**1,441,737**	**100.00**	**1,435,779**	**100.00**
	MARGINS:								
9	New Equipment	79,434	10.32			96,370	12.28	62,750	8.78
10	Used Equipment	17,092	6.86			25,329	11.59	16,064	6.05
11	*Total New and Used*	96,526	9.47			121,699	12.13	78,814	8.04
12	Repair Parts	84,108	30.18			94,176	31.99	74,941	28.92
13	Service Labor	31,317	33.01			28,204	33.07	34,514	32.45
14	All Other Lines	8,279	13.97			7,951	16.77	9,506	12.71
15	Rental-Lease Income	4,768	40.86			5,503	47.25	6,994	44.67
16	**TOTAL OPERATING MARGIN**	**224,998**	**15.38**			**257,533**	**17.86**	**204,769**	**14.26**
	EXPENSES:								
17	Salaries-Salesmen	17,224	1.17			15,178	1.05	18,918	1.31
18	Travel and Training	3,133	.22			2,860	.20	3,349	.24
19	Advertising and Promotion	9,266	.63			8,079	.56	10,344	.72
20	After Sales & Warranty Exp.	8,513	.58			6,346	.44	10,914	.76
21	**TOTAL SALES EXPENSES**	**38,136**	**2.60**			**32,463**	**2.25**	**43,525**	**3.03**
22	Salaries - Officers, Owners	28,850	1.97			31,669	2.19	27,752	1.93
23	Salaries - Partsmen	19,772	1.35			19,461	1.35	20,560	1.43
24	Salaries - Office and Others	16,014	1.10			14,995	1.04	17,929	1.25
25	Occupancy Expense	23,997	1.64			19,874	1.38	30,248	2.11
26	Heat, Light, Power, Water	5,363	.36			4,715	.33	5,910	.41
27	Telephone and Telegraph	4,114	.29			3,492	.24	4,791	.33
28	Office & Store Exp., and Postage	7,510	.51			6,416	.45	8,190	.57
29	Shop Supplies and Expense	6,833	.47			6,356	.44	7,192	.50
30	Car and Truck Expense	15,034	1.02			13,082	.90	16,686	1.17
31	Taxes - Except Payroll & Real Estate	3,219	.22			3,381	.24	3,302	.23
32	Employee Benefits	18,502	1.27			16,408	1.14	21,127	1.47
33	Dues and Subscriptions	667	.04			647	.04	712	.05
34	Insurance - Except Group & R.E.	8,898	.61			8,914	.62	9,222	.64
35	Legal and Auditing	2,701	.19			2,329	.16	3,136	.22
36	Interest & Bank Chgs - Except R.E.	19,805	1.35			10,184	.71	32,256	2.24
37	Depreciation — Except R.E.	9,047	.62			7,069	.49	11,201	.78
38	Bad Debts	2,136	.14			2,550	.17	1,744	.13
39	Contributions and Miscellaneous	2,130	.15			2,323	.17	1,817	.12
40	**TOTAL EXPENSES**	**232,728**	**15.90**			**206,328**	**14.31**	**267,300**	**18.61**
41	**NET PROFIT ON SALES**	**(7,730)**	**(.52)**			**51,205**	**3.55**	**(62,531)**	**(4.35)**
42	Other Income	37,882	2.58			44,920	3.11	28,921	2.01
43	**NET OPERATING PROFIT OR LOSS (Before Income Taxes)**	**30,152**	**2.06**			**96,125**	**6.66**	**(33,610)**	**(2.34)**

GROUP "C" OPERATING AVERAGES
DEALERS WITH SALES $2,000,000 TO $3,000,000

LINE NO.	ACCOUNT DESCRIPTION	AVERAGE Amount	AVERAGE % Sales	YOUR FIGURES Amount	YOUR FIGURES % Sales	HIGH PROFIT Amount	HIGH PROFIT % Sales	LOW PROFIT Amount	LOW PROFIT % Sales
	SALES:								
1	New Equipment	$1,276,682	52.04			$1,263,585	51.63	$1,213,021	50.23
2	Used Equipment	443,858	18.09			432,590	17.68	421,674	17.47
3	Total New and Used	1,720,540	70.13			1,696,175	69.31	1,634,695	67.70
4	Repair Parts	419,949	17.12			433,159	17.70	422,440	17.49
5	Service Labor	157,466	6.42			148,773	6.08	166,335	6.89
6	All Other Lines	138,336	5.64			156,584	6.40	162,939	6.75
7	Rental-Lease Income	16,890	.69			12,437	.51	28,141	1.17
8	**TOTAL SALES**	2,453,181	100.00			2,447,128	100.00	2,414,550	100.00
	MARGINS:								
9	New Equipment	125,844	9.86			161,820	12.81	90,937	7.50
10	Used Equipment	31,560	7.11			42,016	9.71	18,972	4.50
11	Total New and Used	157,404	9.15			203,836	12.02	109,909	6.72
12	Repair Parts	128,892	30.69			131,905	30.45	129,858	30.74
13	Service Labor	57,940	36.80			56,519	37.99	56,969	34.25
14	All Other Lines	19,412	14.03			28,965	18.50	22,094	13.56
15	Rental-Lease Income	6,144	36.38			4,520	36.34	6,955	24.71
16	**TOTAL OPERATING MARGIN**	369,792	15.07			425,745	17.40	325,785	13.49
	EXPENSES:								
17	Salaries-Salesmen	35,988	1.46			33,485	1.36	35,052	1.45
18	Travel and Training	4,998	.21			4,712	.20	4,780	.19
19	Advertising and Promotion	14,417	.58			12,801	.52	14,280	.60
20	After Sales & Warranty Exp.	15,479	.63			13,620	.56	17,308	.71
21	**TOTAL SALES EXPENSES**	70,882	2.88			64,618	2.64	71,420	2.95
22	Salaries - Officers, Owners	45,156	1.85			45,143	1.84	44,069	1.83
23	Salaries - Partsmen	32,829	1.33			31,376	1.28	35,876	1.48
24	Salaries - Office and Others	28,487	1.16			32,328	1.32	31,215	1.30
25	Occupancy Expense	37,327	1.53			31,911	1.31	40,076	1.66
26	Heat, Light, Power, Water	7,607	.31			6,961	.28	8,346	.34
27	Telephone and Telegraph	6,008	.24			5,463	.23	6,923	.29
28	Office & Store Exp., and Postage	12,206	.50			10,194	.41	13,698	.57
29	Shop Supplies and Expense	9,484	.39			8,826	.36	10,537	.43
30	Car and Truck Expense	22,739	.92			20,568	.84	26,616	1.10
31	Taxes - Except Payroll & Real Estate	4,958	.20			4,559	.19	5,495	.23
32	Employee Benefits	31,424	1.29			31,968	1.31	31,725	1.32
33	Dues and Subscriptions	925	.03			828	.03	1,069	.04
34	Insurance - Except Group & R.E.	12,302	.50			11,854	.48	13,440	.56
35	Legal and Auditing	3,492	.15			3,635	.15	4,515	.18
36	Interest & Bank Chgs - Except R.E.	25,848	1.05			13,011	.53	42,712	1.77
37	Depreciation — Except R.E.	11,371	.46			9,936	.41	11,801	.49
38	Bad Debts	3,498	.15			3,583	.15	3,747	.16
39	Contributions and Miscellaneous	3,588	.14			3,362	.13	4,806	.20
40	**TOTAL EXPENSES**	370,131	15.08			340,124	13.89	408,086	16.90
41	**NET PROFIT ON SALES**	(339)	(.01)			85,621	3.50	(82,301)	(3.40)
42	Other Income	55,388	2.25			62,602	2.55	48,806	2.02
43	**NET OPERATING PROFIT OR LOSS (Before Income Taxes)**	55,049	2.24			148,223	6.05	(33,495)	(1.38)

GROUP "D" OPERATING AVERAGES
DEALERS WITH SALES OVER $3,000,000

LINE NO.	ACCOUNT DESCRIPTION	AVERAGE Amount	AVERAGE % Sales	YOUR FIGURES Amount	YOUR FIGURES % Sales	HIGH PROFIT Amount	HIGH PROFIT % Sales	LOW PROFIT Amount	LOW PROFIT % Sales
	SALES:								
1	New Equipment	$2,521,384	52.38			$2,832,623	52.58	$2,243,289	52.34
2	Used Equipment	880,928	18.30			830,333	15.42	897,309	20.94
3	*Total New and Used*	3,402,312	70.68			3,662,956	68.00	3,140,598	73.28
4	Repair Parts	786,132	16.33			1,005,508	18.67	652,051	15.21
5	Service Labor	328,713	6.83			412,347	7.65	280,120	6.54
6	All Other Lines	219,975	4.57			250,994	4.66	144,454	3.37
7	Rental-Lease Income	76,087	1.59			54,618	1.02	68,322	1.60
8	**TOTAL SALES**	4,813,219	100.00			5,386,423	100.00	4,285,545	100.00
	MARGINS:								
9	New Equipment	249,452	9.89			328,829	11.61	211,616	9.43
10	Used Equipment	61,044	6.93			107,225	12.91	17,505	1.95
11	*Total New and Used*	310,496	9.13			436,054	11.90	229,121	7.30
12	Repair Parts	243,917	31.03			318,222	31.65	193,035	29.60
13	Service Labor	131,545	40.02			171,220	41.52	108,258	38.65
14	All Other Lines	34,871	15.85			44,129	17.58	23,058	15.96
15	Rental-Lease Income	19,508	25.64			19,568	35.83	16,836	24.64
16	**TOTAL OPERATING MARGIN**	740,337	15.38			989,193	18.36	570,308	13.31
	EXPENSES:								
17	Salaries-Salesmen	68,915	1.43			78,633	1.45	64,712	1.51
18	Travel and Training	9,770	.20			12,493	.24	8,865	.20
19	Advertising and Promotion	24,646	.51			26,627	.49	24,946	.58
20	After Sales & Warranty Exp.	38,348	.80			38,485	.72	40,347	.95
21	**TOTAL SALES EXPENSES**	141,679	2.94			156,238	2.90	138,870	3.24
22	Salaries - Officers, Owners	78,290	1.63			81,080	1.50	66,389	1.54
23	Salaries - Partsmen	56,321	1.17			68,793	1.28	46,777	1.10
24	Salaries - Office and Others	47,819	.99			51,072	.95	40,591	.94
25	Occupancy Expense	61,782	1.28			70,308	1.30	63,275	1.48
26	Heat, Light, Power, Water	10,955	.23			11,193	.21	10,690	.25
27	Telephone and Telegraph	10,193	.21			11,423	.21	9,578	.22
28	Office & Store Exp., and Postage	22,062	.46			23,423	.44	22,450	.53
29	Shop Supplies and Expense	18,534	.39			23,800	.44	15,560	.36
30	Car and Truck Expense	42,866	.89			51,699	.96	39,174	.91
31	Taxes - Except Payroll & Real Estate	8,318	.17			9,236	.17	7,685	.18
32	Employee Benefits	60,583	1.26			72,561	1.35	53,136	1.24
33	Dues and Subscriptions	1,374	.03			1,608	.03	1,209	.03
34	Insurance - Except Group & R.E.	21,010	.43			22,381	.41	21,072	.49
35	Legal and Auditing	5,553	.12			7,521	.14	5,617	.13
36	Interest & Bank Chgs - Except R.E.	44,005	.91			39,090	.73	60,489	1.42
37	Depreciation — Except R.E.	24,307	.51			27,057	.50	23,031	.53
38	Bad Debts	7,365	.15			6,663	.12	5,091	.12
39	Contributions and Miscellaneous	9,166	.19			7,925	.15	6,516	.15
40	**TOTAL EXPENSES**	672,182	13.96			743,071	13.79	637,200	14.86
41	**NET PROFIT ON SALES**	68,155	1.42			246,122	4.57	(66,892)	(1.56)
42	Other Income	114,200	2.37			173,511	3.22	78,826	1.83
43	**NET OPERATING PROFIT OR LOSS** (Before Income Taxes)	182,355	3.79			419,633	7.79	11,934	.27

ASSETS VS. DEBT

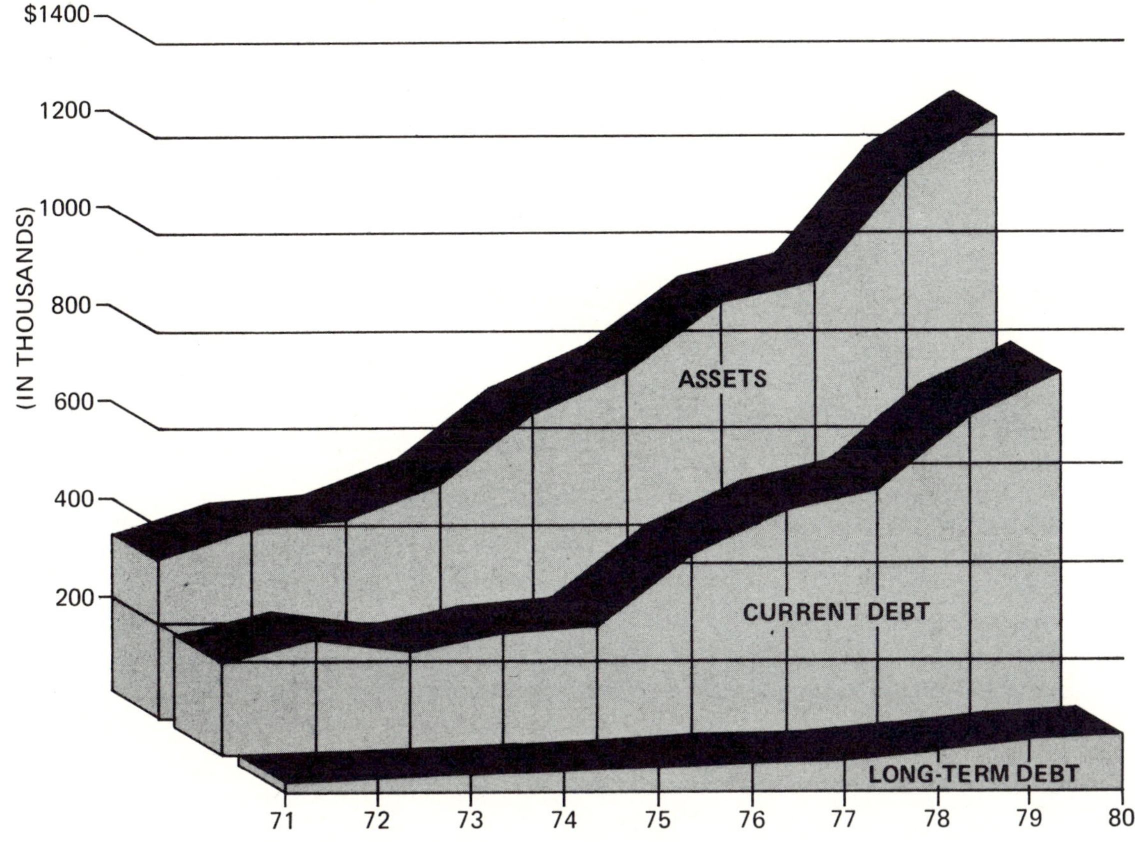

ASSETS VS. DEBT
(In Thousands)

	1971	1972	1973	1974	1975	1976	1977	1978	1979	1980
Assets	316.1	380.1	399.8	465.9	617.5	688.8	860.5	895.7	1107.8	1238.7
Short-Term Debt	186.7	225.5	205.1	243.5	353.7	407.0	489.3	537.4	684.3	785.0
Long-Term Debt	26.7	32.0	36.9	38.0	47.9	60.8	62.3	84.4	99.0	116.5

CAPITALIZATION
(Percent of Total)

	1971	1972	1973	1974	1975	1976	1977	1978	1979	1980
Outside Capital	67.5	67.7	60.5	60.4	64.6	67.9	64.1	69.4	70.7	72.8
Inside Capital	32.5	32.3	39.5	39.6	35.4	32.1	35.9	30.6	29.3	27.2

BALANCE SHEET AND INVENTORY STUDIES

ALL DEALERS

BALANCE SHEET ACCOUNTS	AVERAGE ALL VOLUME GROUPS			YOUR BALANCE SHEET FIGURES		HIGH PROFIT DEALERS			LOW PROFIT DEALERS		
Assets	Amount	% Total Assets	Turn Over	Amount	% Total Assets	Amount	% Total Assets	Turn Over	Amount	% Total Assets	Turn Over
Cash and Securities	$ 46,669	3.76				$ 80,908	5.61		$ 22,853	2.27	
Accounts and Notes Receivable	135,745	10.96				193,032	13.39		83,779	8.33	
TOTAL	182,414	14.72				273,940	19.00		106,632	10.60	
Inventories:											
New Equipment	549,851	44.39	1.92			605,737	42.02	2.04	470,572	46.81	1.52
Used Equipment	148,047	11.95	2.51			144,308	10.01	2.62	140,047	13.92	1.92
Repair Parts	148,501	11.99	1.84			162,773	11.29	2.12	123,278	12.27	1.57
Other Lines	40,815	3.30	2.19			40,499	2.81	2.59	39,561	3.93	1.69
Rental or Lease	13,908	1.12	1.34			17,449	1.21	1.03	9,674	.96	1.67
TOTAL INVENTORIES	901,122	72.75	2.01			970,766	67.34	2.14	783,132	77.89	1.61
TOTAL CURRENT ASSETS	1,083,536	87.47				1,244,706	86.34		889,764	88.49	
Fixed Assets	117,448	9.48				135,211	9.38		89,248	8.88	
Finance Reserve and Other Assets	37,693	3.05				61,553	4.28		26,390	2.63	
TOTAL ASSETS	1,238,677	100.00				1,441,470	100.00		1,005,402	100.00	
Liabilities and Net Worth											
Current Accts. Payable & Accruals	134,685					158,345			88,159		
Floor Plan Notes Payable	545,363					564,303			483,816		
Notes Payable — Bank and Other	104,968					72,215			134,032		
Mortgages and Other Long Term	116,482					119,466			127,332		
TOTAL LIABILITIES	901,498					914,329			833,339		
Net Worth	337,179	27.22				527,141	36.57		172,063	17.11	
TOTAL LIABILITIES AND NET WORTH	1,238,677					1,441,470			1,005,402		

ALL DEALERS — REGIONAL

BALANCE SHEET ACCOUNTS	EASTERN REGION			YOUR BALANCE SHEET FIGURES		CENTRAL REGION			WESTERN REGION		
Assets	Amount	% Total Assets	Turn Over	Amount	% Total Assets	Amount	% Total Assets	Turn Over	Amount	% Total Assets	Turn Over
Cash and Securities	48,467	4.53				40,074	3.37		56,546	3.91	
Accounts and Notes Receivable	136,390	12.76				108,180	9.13		181,753	12.58	
TOTAL	184,857	17.29				148,254	12.50		238,299	16.49	
Inventories:											
New Equipment	445,483	41.68	1.82			558,057	47.05	1.85	607,792	42.09	2.09
Used Equipment	106,614	9.98	2.43			152,958	12.90	2.59	168,265	11.65	2.42
Repair Parts	133,292	12.47	2.02			137,099	11.55	1.78	178,174	12.33	1.83
Other Lines	58,233	5.45	2.18			39,149	3.31	1.81	31,643	2.20	3.01
Rental or Lease	13,982	1.31	1.87			13,710	1.15	.80	14,191	.98	1.86
TOTAL INVENTORIES	757,604	70.89	1.97			900,973	75.96	1.95	1,000,065	69.25	2.13
TOTAL CURRENT ASSETS	942,461	88.18				1,049,227	88.46		1,238,364	85.74	
Fixed Assets	97,969	9.17				104,059	8.78		153,405	10.62	
Finance Reserve and Other Assets	28,308	2.65				32,729	2.76		52,510	3.64	
TOTAL ASSETS	1,068,738	100.00				1,186,015	100.00		1,444,279	100.00	
Liabilities and Net Worth											
Current Accts. Payable & Accruals	154,328					125,589			136,505		
Floor Plan Notes Payable	419,724					547,692			627,836		
Notes Payable — Bank and Other	82,761					104,315			121,338		
Mortgages and Other Long Term	106,837					105,267			142,012		
TOTAL LIABILITIES	763,650					882,863			1,027,691		
Net Worth	305,088	28.55				303,152	25.56		416,588	28.84	
TOTAL LIABILITIES AND NET WORTH	1,068,738					1,186,015			1,444,279		

GROUP "A" — DEALERS WITH SALES TO $1,000,000

BALANCE SHEET ACCOUNTS Assets	THIS VOLUME GROUP Amount	% Total Assets	Turn Over	YOUR BALANCE SHEET FIGURES Amount	% Total Assets	HIGH PROFIT DEALERS Amount	% Total Assets	Turn Over	LOW PROFIT DEALERS Amount	% Total Assets	Turn Over
Cash and Securities	$ 21,715	4.05				$ 37,716	6.52		$ 10,026	1.76	
Accounts and Notes Receivable	42,359	7.91				54,997	9.51		42,649	7.50	
TOTAL	64,074	11.96				92,713	16.03		52,675	9.26	
Inventories:											
New Equipment	257,078	47.99	1.27			255,472	44.17	1.49	283,977	49.97	.95
Used Equipment	59,717	11.15	1.64			62,764	10.85	1.82	72,695	12.80	1.20
Repair Parts	72,344	13.51	1.52			64,426	11.14	1.84	78,963	13.89	1.27
Other Lines	18,511	3.45	1.29			28,232	4.88	.98	19,342	3.40	.98
Rental or Lease	7,041	1.32	.96			7,153	1.24	1.43	3,082	.55	.24
TOTAL INVENTORIES	414,691	77.42	1.36			418,047	72.28	1.56	458,059	80.61	1.04
TOTAL CURRENT ASSETS	478,765	89.38				510,760	88.31		510,734	89.87	
Fixed Assets	47,622	8.89				54,467	9.42		50,661	8.91	
Finance Reserve and Other Assets	9,253	1.73				13,120	2.27		6,905	1.22	
TOTAL ASSETS	535,640	100.00				578,347	100.00		568,300	100.00	
Liabilities and Net Worth											
Current Accts. Payable & Accruals	41,790					37,446			43,894		
Floor Plan Notes Payable	247,637					252,067			266,831		
Notes Payable — Bank and Other	46,451					30,370			69,612		
Mortgages and Other Long Term	62,223					45,186			93,690		
TOTAL LIABILITIES	398,101					365,069			474,027		
Net Worth	137,539	25.68				213,278	36.88		94,273	16.59	
TOTAL LIABILITIES AND NET WORTH	535,640					578,347			568,300		

GROUP "B" — DEALERS WITH SALES $1,000,000 TO $2,000,000

BALANCE SHEET ACCOUNTS Assets	THIS VOLUME GROUP Amount	% Total Assets	Turn Over	YOUR BALANCE SHEET FIGURES Amount	% Total Assets	HIGH PROFIT DEALERS Amount	% Total Assets	Turn Over	LOW PROFIT DEALERS Amount	% Total Assets	Turn Over
Cash and Securities	32,302	3.58				52,733	6.16		18,521	1.87	
Accounts and Notes Receivable	77,224	8.56				89,749	10.50		77,617	7.86	
TOTAL	109,526	12.14				142,482	16.66		96,138	9.73	
Inventories:											
New Equipment	436,165	48.38	1.58			386,747	45.23	1.78	469,130	47.49	1.39
Used Equipment	107,635	11.93	2.16			87,463	10.23	2.21	137,537	13.93	1.81
Repair Parts	109,918	12.19	1.77			101,928	11.92	1.96	118,510	12.00	1.55
Other Lines	30,019	3.33	1.70			18,406	2.15	2.14	42,243	4.27	1.55
Rental or Lease	7,262	.81	.95			6,396	.75	.96	8,982	.91	.96
TOTAL INVENTORIES	690,999	76.64	1.70			600,940	70.28	1.88	776,402	78.60	1.49
TOTAL CURRENT ASSETS	800,525	88.78				743,422	86.94		872,540	88.33	
Fixed Assets	82,148	9.11				91,592	10.71		89,488	9.06	
Finance Reserve and Other Assets	18,991	2.11				20,075	2.35		25,713	2.61	
TOTAL ASSETS	901,664	100.00				855,089	100.00		987,741	100.00	
Liabilities and Net Worth											
Current Accts. Payable & Accruals	81,567					75,654			76,501		
Floor Plan Notes Payable	434,209					365,635			500,303		
Notes Payable — Bank and Other	75,212					51,010			123,232		
Mortgages and Other Long Term	92,465					70,780			127,873		
TOTAL LIABILITIES	683,453					563,079			827,909		
Net Worth	218,211	24.20				292,010	34.15		159,832	16.18	
TOTAL LIABILITIES AND NET WORTH	901,664					855,089			987,741		

BALANCE SHEET AND INVENTORY STUDIES

GROUP "C" — DEALERS WITH SALES $2,000,000 TO $3,000,000

BALANCE SHEET ACCOUNTS	THIS VOLUME GROUP			YOUR BALANCE SHEET FIGURES			HIGH PROFIT DEALERS			LOW PROFIT DEALERS		
Assets	Amount	% Total Assets	Turn Over	Amount	% Total Assets		Amount	% Total Assets	Turn Over	Amount	% Total Assets	Turn Over
Cash and Securities	$ 52,424	3.93		$			75,904	5.74	$	47,012	3.35	
Accounts and Notes Receivable	144,515	10.84					153,136	11.60		138,918	9.90	
TOTAL	**196,939**	**14.77**					**229,040**	**17.34**		**185,930**	**13.25**	
Inventories:												
New Equipment	575,606	43.18	2.00				563,553	42.68	1.96	594,502	42.40	1.89
Used Equipment	170,164	12.77	2.42				156,382	11.84	2.50	184,665	13.16	2.18
Repair Parts	172,183	12.92	1.69				157,607	11.94	1.91	179,374	12.79	1.63
Other Lines	53,068	3.98	2.24				47,728	3.61	2.67	72,751	5.19	1.94
Rental or Lease	9,880	.74	1.09				6,327	.48	1.25	14,790	1.06	1.43
TOTAL INVENTORIES	**980,901**	**73.59**	**2.02**				**931,597**	**70.55**	**2.07**	**1,046,082**	**74.60**	**1.89**
TOTAL CURRENT ASSETS	**1,177,840**	**88.36**					**1,160,637**	**87.89**		**1,232,012**	**87.85**	
Fixed Assets	114,785	8.61					111,428	8.44		117,950	8.41	
Finance Reserve and Other Assets	40,310	3.03					48,420	3.67		52,416	3.74	
TOTAL ASSETS	**1,332,935**	**100.00**					**1,320,485**	**100.00**		**1,402,378**	**100.00**	
Liabilities and Net Worth												
Current Accts. Payable & Accruals	159,009						163,495			143,822		
Floor Plan Notes Payable	563,086						501,631			611,195		
Notes Payable — Bank and Other	121,072						52,050			219,752		
Mortgages and Other Long Term	124,749						106,086			159,636		
TOTAL LIABILITIES	**967,916**						**823,262**			**1,134,405**		
Net Worth	365,019	27.38					497,223	37.65		267,973	19.11	
TOTAL LIABILITIES AND NET WORTH	**1,332,935**						**1,320,485**			**1,402,378**		

GROUP "D" — DEALERS WITH SALES OVER $3,000,000

BALANCE SHEET ACCOUNTS	THIS VOLUME GROUP			YOUR BALANCE SHEET FIGURES			HIGH PROFIT DEALERS			LOW PROFIT DEALERS		
Assets	Amount	% Total Assets	Turn Over	Amount	% Total Assets		Amount	% Total Assets	Turn Over	Amount	% Total Assets	Turn Over
Cash and Securities	88,160	3.71					150,892	5.54		37,968	1.84	
Accounts and Notes Receivable	311,344	13.12					421,157	15.47		214,135	10.40	
TOTAL	**399,504**	**16.83**					**572,049**	**21.01**		**252,103**	**12.24**	
Inventories:												
New Equipment	995,689	41.94	2.28				1,113,219	40.90	2.25	949,403	46.11	2.14
Used Equipment	277,195	11.68	2.96				245,673	9.03	2.94	317,437	15.42	2.77
Repair Parts	260,832	10.99	2.08				300,902	11.06	2.28	224,976	10.93	2.04
Other Lines	67,499	2.84	2.74				56,471	2.07	3.66	61,201	2.97	1.98
Rental or Lease	35,180	1.48	1.61				28,083	1.03	1.25	19,907	.97	2.59
TOTAL INVENTORIES	**1,636,395**	**68.93**	**2.37**				**1,744,348**	**64.09**	**2.38**	**1,572,924**	**76.40**	**2.25**
TOTAL CURRENT ASSETS	**2,035,899**	**85.76**					**2,316,397**	**85.10**		**1,825,027**	**88.64**	
Fixed Assets	245,284	10.34					275,812	10.14		179,848	8.73	
Finance Reserve and Other Assets	92,565	3.90					129,545	4.76		54,012	2.63	
TOTAL ASSETS	**2,373,748**	**100.00**					**2,721,754**	**100.00**		**2,058,887**	**100.00**	
Liabilities and Net Worth												
Current Accts. Payable & Accruals	285,613						274,669			259,469		
Floor Plan Notes Payable	1,000,618						1,030,575			1,023,384		
Notes Payable — Bank and Other	193,591						125,527			184,600		
Mortgages and Other Long Term	200,027						235,662			177,092		
TOTAL LIABILITIES	**1,679,849**						**1,666,433**			**1,644,545**		
Net Worth	693,899	29.23					1,055,321	38.77		414,342	20.12	
TOTAL LIABILITIES AND NET WORTH	**2,373,748**						**2,721,754**			**2,058,887**		

OPERATING AND FINANCIAL RATIOS

ALL DEALERS

DESCRIPTIONS	AVERAGE ALL VOLUME GROUPS	YOUR RATIOS	HIGH PROFIT DEALERS	LOW PROFIT DEALERS
Current Ratio	1.38 TO 1		1.57 TO 1	1.26 TO 1
Ownership Equity	27.22 %		36.57 %	17.11 %
Receivables Turnover	16.59 TIMES		13.87 TIMES	18.51 TIMES
Net Working Capital Turnover	7.54 TIMES		5.95 TIMES	8.44 TIMES
Inventory to Net Working Capital	3.02 TO 1		2.16 TO 1	4.26 TO 1
Turnover of Total Assets	1.82 TIMES		1.86 TIMES	1.54 TIMES
Return on Total Assets	5.23 %		12.71 %	(2.86 %)
Return on Net Worth	19.19 %		34.77 %	(16.74 %)

ALL DEALERS — REGIONAL

	EASTERN REGION		CENTRAL REGION	WESTERN REGION
Current Ratio	1.43 TO 1		1.35 TO 1	1.40 TO 1
Ownership Equity	28.55 %		25.56 %	28.84 %
Receivables Turnover	13.91 TIMES		19.99 TIMES	14.57 TIMES
Net Working Capital Turnover	6.64 TIMES		7.96 TIMES	7.51 TIMES
Inventory to Net Working Capital	2.65 TO 1		3.32 TO 1	2.84 TO 1
Turnover of Total Assets	1.78 TIMES		1.82 TIMES	1.83 TIMES
Return on Total Assets	4.47 %		4.26 %	6.94 %
Return on Net Worth	15.67 %		16.67 %	24.06 %

GROUP "A" — DEALERS WITH SALES TO $1,000,000

DESCRIPTIONS	THIS VOLUME GROUP	YOUR RATIOS	HIGH PROFIT DEALERS	LOW PROFIT DEALERS
Current Ratio	1.43 TO 1		1.60 TO 1	1.34 TO 1
Ownership Equity	25.68 %		36.88 %	16.59 %
Receivables Turnover	17.02 TIMES		15.42 TIMES	14.24 TIMES
Net Working Capital Turnover	5.04 TIMES		4.44 TIMES	4.66 TIMES
Inventory to Net Working Capital	2.90 TO 1		2.19 TO 1	3.51 TO 1
Turnover of Total Assets	1.35 TIMES		1.47 TIMES	1.07 TIMES
Return on Total Assets	2.02 %		9.87 %	(5.51 %)
Return on Net Worth	7.87 %		26.75 %	(33.22 %)

GROUP "B" — DEALERS WITH SALES $1,000,000 TO $2,000,000

	THIS VOLUME GROUP		HIGH PROFIT DEALERS	LOW PROFIT DEALERS
Current Ratio	1.35 TO 1		1.51 TO 1	1.25 TO 1
Ownership Equity	24.20 %		34.15 %	16.18 %
Receivables Turnover	18.95 TIMES		16.06 TIMES	18.50 TIMES
Net Working Capital Turnover	6.98 TIMES		5.74 TIMES	8.32 TIMES
Inventory to Net Working Capital	3.30 TO 1		2.39 TO 1	4.50 TO 1
Turnover of Total Assets	1.62 TIMES		1.69 TIMES	1.45 TIMES
Return on Total Assets	3.34 %		11.24 %	(3.40 %)
Return on Net Worth	13.82 %		32.92 %	(21.02 %)

OPERATING AND FINANCIAL RATIOS

GROUP "C" — DEALERS WITH SALES $2,000,000 TO $3,000,000

DESCRIPTIONS	THIS VOLUME GROUP	YOUR RATIOS	HIGH PROFIT DEALERS	LOW PROFIT DEALERS
Current Ratio	1.40 TO 1		1.62 TO 1	1.26 TO 1
Ownership Equity	27.38 %		37.65 %	19.11 %
Receivables Turnover	16.98 TIMES		15.98 TIMES	17.38 TIMES
Net Working Capital Turnover	7.33 TIMES		5.52 TIMES	9.39 TIMES
Inventory to Net Working Capital	2.93 TO 1		2.10 TO 1	4.07 TO 1
Turnover of Total Assets	1.84 TIMES		1.85 TIMES	1.72 TIMES
Return on Total Assets	4.13 %		11.22 %	(2.38 %)
Return on Net Worth	15.08 %		29.81 %	(12.49 %)

GROUP "D" — DEALERS WITH SALES OVER $3,000,000

DESCRIPTIONS	THIS VOLUME GROUP	YOUR RATIOS	HIGH PROFIT DEALERS	LOW PROFIT DEALERS
Current Ratio	1.38 TO 1		1.62 TO 1	1.24 TO 1
Ownership Equity	29.23 %		38.77 %	20.12 %
Receivables Turnover	15.46 TIMES		12.79 TIMES	20.01 TIMES
Net Working Capital Turnover	8.66 TIMES		6.08 TIMES	11.99 TIMES
Inventory to Net Working Capital	2.94 TO 1		1.97 TO 1	4.40 To 1
Turnover of Total Assets	2.03 TIMES		1.98 TIMES	2.08 TIMES
Return on Total Assets	7.68 %		15.42 %	.58 %
Return on Net Worth	26.28 %		39.76 %	2.88 %

PERSONNEL STUDIES

ALL DEALERS AND REGIONAL

DESCRIPTIONS	ALL VOLUME GROUPS	YOUR RATIOS	EASTERN REGION	CENTRAL REGION	WESTERN REGION
Number of Employees per Dealer (incl. owners)	13.66		16.26	12.34	13.56
Total Sales per Employee	$164,884		$162,839	$175,240	$139,928
Total Assets per Employee	$ 90,679		$ 88,824	$ 96,111	$ 78,816
Number of Salesmen (including Owners)	2.96		3.41	2.75	2.91
New and Used Sales per Salesman	$532,397		$546,752	$569,385	$405,851
Number of Partsmen	2.16		2.56	1.95	2.14
Parts Sales per Partsman	$182,883		$184,472	$180,724	$180,916
Number of Servicemen	6.21		7.53	5.52	6.21
Service Sales per Serviceman	$ 24,135		$ 21,937	$ 25,876	$ 23,327
Number of Office Employees	1.66		2.00	1.46	1.70
Number of Other Employees	.67		.76	.66	.60

VOLUME GROUPS

DESCRIPTIONS	GROUP "A"	YOUR RATIOS	GROUP "B"	GROUP "C"	GROUP "D"
Number of Employees per Dealer (incl. owners)	6.99		10.20	14.40	23.71
Total Sales per Employee	$103,116		$143,462	$170,360	$203,004
Total Assets per Employee	$ 76,629		$ 88,399	$ 92,565	$100,116
Number of Salesmen (including Owners)	1.77		2.38	3.17	4.63
New and Used Sales per Salesman	$267,716		$428,064	$542,757	$734,841
Number of Partsmen	1.16		1.62	2.24	3.71
Parts Sales per Partsman	$138,510		$172,037	$187,478	$211,895
Number of Servicemen	2.74		4.41	6.61	11.44
Service Sales per Serviceman	$ 17,973		$ 21,512	$ 23,822	$ 28,734
Number of Office Employees	1.02		1.34	1.72	2.61
Number of Other Employees	.30		.45	.66	1.33

INDUSTRIAL COST OF DOING BUSINESS STUDY
ALL DEALERS OPERATING AVERAGES

LINE NO.		AVERAGE Amount	AVERAGE % Sales	YOUR FIGURES Amount	YOUR FIGURES % Sales	HIGH PROFIT DEALERS Amount	HIGH PROFIT DEALERS % Sales	LOW PROFIT DEALERS Amount	LOW PROFIT DEALERS % Sales
	SALES:								
1	New Equipment	$1,289,467	46.83			$1,543,935	46.36	$ 837,742	44.23
2	Used Equipment	363,851	13.21			408,911	12.28	227,973	12.03
3	*Total New and Used*	1,653,318	60.04			1,952,846	58.64	1,065,715	56.26
4	Repair Parts	533,215	19.37			616,015	18.50	422,781	22.33
5	Service Labor	191,403	6.95			216,603	6.51	138,145	7.29
6	All Other Lines	192,772	7.00			345,120	10.36	142,648	7.53
7	Rental-Lease Income	182,546	6.64			199,169	5.99	124,657	6.59
8	**TOTAL SALES**	2,753,254	100.00			3,329,753	100.00	1,893,946	100.00
	MARGINS								
9	New Equipment	157,910	12.25			222,350	14.40	97,925	11.69
10	Used Equipment	27,612	7.59			65,113	15.92	24,731	10.85
11	*Total New and Used*	185,522	11.22			287,463	14.72	122,656	11.51
12	Repair Parts	178,919	33.55			221,305	35.93	153,932	36.41
13	Service Labor	65,704	34.33			76,124	35.14	21,595	15.63
14	All Other Lines	30,437	15.79			65,434	18.96	16,616	11.65
15	Rental-Lease Income	35,291	19.33			58,028	29.14	3,043	2.44
16	**TOTAL OPERATING MARGIN**	495,873	18.01			708,354	21.27	317,842	16.78
	EXPENSES:								
17	Salaries-Salesmen	46,271	1.68			62,632	1.88	36,786	1.94
18	Travel and Training	7,829	.28			10,984	.33	5,827	.30
19	Advertising and Promotion	18,045	.66			25,359	.76	13,871	.74
20	After Sales & Warranty Exp.	13,768	.50			16,388	.49	6,608	.35
21	**TOTAL SALES EXPENSES**	85,913	3.12			115,363	3.46	63,092	3.33
22	Salaries - Officers, Owners	54,111	1.96			61,039	1.83	37,507	1.98
23	Salaries - Partsmen	39,900	1.45			50,195	1.51	29,815	1.57
24	Salaries - Office and Others	38,171	1.39			51,443	1.55	31,145	1.65
25	Occupancy Expense	44,693	1.62			46,282	1.39	33,862	1.78
26	Heat, Light, Power, Water	7,388	.27			8,055	.24	7,273	.39
27	Telephone and Telegraph	7,742	.28			9,511	.28	6,688	.35
28	Office & Store Exp., and Postage	13,833	.50			14,379	.43	11,005	.58
29	Shop Supplies and Expense	17,809	.65			17,904	.54	10,130	.54
30	Car and Truck Expense	31,991	1.16			39,504	1.19	21,614	1.14
31	Taxes - Except Payroll & Real Estate	7,511	.27			9,758	.29	5,784	.30
32	Employee Benefits	38,038	1.38			55,520	1.67	29,228	1.55
33	Dues and Subscriptions	1,377	.05			1,689	.05	1,368	.07
34	Insurance - Except Group & R.E.	17,035	.62			20,047	.60	13,121	.69
35	Legal and Auditing	6,121	.23			8,835	.27	4,422	.23
36	Interest & Bank Chgs - Except R.E.	41,740	1.51			48,980	1.47	52,859	2.80
37	Depreciation	13,563	.49			16,958	.51	11,734	.62
38	Bad Debts	6,800	.25			9,977	.30	3,850	.20
39	Contributions and Miscellaneous	4,454	.16			4,093	.12	5,094	.27
40	**TOTAL EXPENSES**	478,190	17.36			589,532	17.70	379,591	20.04
41	**NET PROFIT ON SALES**	17,683	.64			118,822	3.57	(61,749)	(3.26)
42	Other Income	75,415	2.73			110,400	3.31	47,988	2.53
43	**NET OPERATING PROFIT OR LOSS (Before Income Taxes)**	93,098	3.37			229,222	6.88	(13,761)	(.73)

OPERATING AND FINANCIAL RATIOS — INDUSTRIAL
ALL DEALERS

	All Dealers	High Profit Dealers	Low Profit Dealers	Your Figures
Current Ratio	1.32 TO 1	1.45 TO 1	1.31 TO 1	
Ownership Equity	26.83 %	33.50 %	21.16 %	
Receivables Turnover	12.70 TIMES	13.83 TIMES	14.76 TIMES	
Net Working Capital Turnover	8.22 TIMES	6.28 TIMES	7.02 TIMES	
Inventory to Net Working Capital	3.27 TO 1	2.62 TO 1	3.58 TO 1	
Turnover of Total Assets	1.69 TIMES	1.70 TIMES	1.51 TIMES	
Return on Total Assets	5.71 %	11.70 %	1.09 % -	
Return on Net Worth	21.29 %	34.92 %	5.19 % -	

BALANCE SHEET AND INVENTORY STUDIES — INDUSTRIAL
ALL DEALERS

BALANCE SHEET ACCOUNTS	AVERAGE ALL VOLUME GROUPS					HIGH PROFIT DEALERS			LOW PROFIT DEALERS		
Assets	Amount	% Total Assets	Turn Over	Amount	% Total Assets	Amount	% Total Assets	Turn Over	Amount	% Total Assets	Turn Over
Cash and Securities	$ 57,203	3.50				$ 86,871	4.43		$ 36,684	2.93	
Accounts and Notes Receivable	216,813	13.30				240,805	12.29		128,350	10.26	
TOTAL	274,016	16.80				327,676	16.72		165,034	13.19	
Inventories:											
New Equipment	619,624	38.01	1.83			810,849	41.37	1.63	547,964	43.80	1.35
Used Equipment	187,090	11.48	1.80			172,890	8.82	1.99	181,527	14.51	1.12
Repair Parts	170,858	10.48	2.07			235,652	12.03	1.67	146,259	11.69	1.84
Other Lines	65,246	4.01	2.49			95,587	4.88	2.93	57,447	4.59	2.19
Rental or Lease	54,039	3.31	2.72			75,642	3.85	1.87	33,713	2.69	3.61
TOTAL INVENTORIES	1,096,857	67.29	1.94			1,390,620	70.95	1.78	966,910	77.28	1.51
TOTAL CURRENT ASSETS	1,370,873	84.09				1,718,296	87.67		1,131,944	90.47	
Fixed Assets	220,780	13.54				186,682	9.53		86,988	6.96	
Finance Reserve and Other Assets	38,484	2.37				54,760	2.80		32,119	2.57	
TOTAL ASSETS	1,630,137	100.00				1,959,738	100.00		1,251,051	100.00	
Liabilities and Net Worth											
Current Accts. Payable & Accruals	184,164					200,274			100,608		
Floor Plan Notes Payable	636,211					825,769			640,802		
Notes Payable — Bank and Other	215,380					162,071			120,634		
Mortgages and Other Long Term	156,998					115,125			124,270		
TOTAL LIABILITIES	1,192,753					1,303,239			986,314		
Net Worth	437,384	26.83				656,499	33.50		264,737	21.16	
TOTAL LIABILITIES AND NET WORTH	1,630,137					1,959,738			1,251,051		

All averages herein are adjusted to nearest dollar, and percentages are adjusted to nearest 1/100th of a percent.

PERSONNEL STUDY — INDUSTRIAL
ALL DEALERS

Number of Employees per Dealer (including Owners & Managers)	16.75
Total Sales per Employee	$164,373
Total Assets per Employee	$ 97,322
Number of Salesmen (including Owners & Managers)	3.57
New and Used Sales per Salesman	$463,115
Number of Partsmen	2.49
Parts Sales per Partsman	$214,143
Number of Servicemen	7.65
Service Sales per Serviceman	$ 25,020
Number of Office Employees	2.11
Number of Other Employees	.93

LAWN & GARDEN COST OF DOING BUSINESS STUDY
ALL DEALERS OPERATING AVERAGES

LINE NO.		AVERAGE		YOUR FIGURES		HIGH PROFIT DEALERS		LOW PROFIT DEALERS	
		Amount	% Sales	Amount	% Sales	Amount	% Sales	Amount	% Sales
	SALES:								
1	New Equipment	$467,195	49.88			$396,639	51.23	$413,516	39.78
2	Used Equipment	112,650	12.03			69,973	9.04	104,811	10.09
3	*Total New and Used*	579,845	61.91			466,612	60.27	518,327	49.87
4	Repair Parts	218,809	23.36			132,199	17.08	322,011	30.98
5	Service Labor	72,763	7.77			43,912	5.67	90,611	8.72
6	All Other Lines	59,452	6.35			126,947	16.40	107,114	10.30
7	Rental-Lease Income	5,683	.61			4,447	.58	1,288	.13
8	**TOTAL SALES**	936,552	100.00			774,117	100.00	1,039,351	100.00
	MARGINS								
9	New Equipment	63,323	13.55			88,222	22.24	38,484	9.31
10	Used Equipment	15,502	13.76			12,755	18.23	(3,131)	(2.98)
11	*Total New and Used*	78,825	13.59			100,977	21.64	35,353	6.82
12	Repair Parts	67,374	30.79			46,147	34.91	96,456	29.95
13	Service Labor	18,597	25.56			13,353	30.41	7,697	8.49
14	All Other Lines	15,634	26.30			38,343	30.20	24,168	22.56
15	Rental-Lease Income	1,965	34.58			1,431	32.18	1,151	89.36
16	**TOTAL OPERATING MARGIN**	182,395	19.48			200,251	25.87	164,825	15.86
	EXPENSES:								
17	Salaries-Salesmen	14,618	1.56			14,341	1.85	14,357	1.38
18	Travel and Training	1,553	.16			1,150	.15	2,948	.28
19	Advertising and Promotion	6,536	.70			4,747	.61	5,711	.55
20	After Sales & Warranty Exp.	5,567	.59			2,811	.36	7,136	.69
21	**TOTAL SALES EXPENSES**	28,274	3.01			23,049	2.97	30,152	2.90
22	Salaries - Officers, Owners	18,018	1.93			5,775	.75	29,990	2.88
23	Salaries - Partsmen	18,841	2.01			14,639	1.89	28,526	2.75
24	Salaries - Office and Others	13,036	1.39			19,887	2.57	10,560	1.01
25	Occupancy Expense	18,873	2.02			13,434	1.73	23,641	2.28
26	Heat, Light, Power, Water	4,156	.44			3,566	.46	6,332	.61
27	Telephone and Telegraph	2,719	.29			2,006	.26	3,745	.36
28	Office & Store Exp., and Postage	5,716	.61			5,352	.70	8,353	.80
29	Shop Supplies and Expense	4,422	.47			7,725	.99	2,016	.19
30	Car and Truck Expense	8,308	.89			8,162	1.06	10,526	1.02
31	Taxes - Except Payroll & Real Estate	2,731	.29			3,279	.42	1,573	.15
32	Employee Benefits	14,366	1.54			15,349	1.98	17,427	1.67
33	Dues and Subscriptions	440	.04			304	.04	1,069	.11
34	Insurance - Except Group & R.E.	5,568	.60			2,864	.37	6,965	.67
35	Legal and Auditing	2,087	.22			2,580	.34	2,655	.25
36	Interest & Bank Chgs - Except R.E.	8,808	.94			7,409	.95	5,074	.49
37	Depreciation	4,962	.53			6,918	.90	5,543	.53
38	Bad Debts	905	.10			291	.03	2,519	.25
39	Contributions and Miscellaneous	3,383	.36			4,730	.62	6,277	.60
40	**TOTAL EXPENSES**	165,613	17.68			147,319	19.03	202,943	19.52
41	**NET PROFIT ON SALES**	16,782	1.79			52,932	6.84	(38,118)	(3.66)
42	**Other Income**	15,692	1.67			25,188	3.25	23,334	2.24
43	**NET OPERATING PROFIT OR LOSS (Before Income Taxes)**	32,474	3.46			78,120	10.09	(14,784)	(1.42)

OPERATING AND FINANCIAL RATIOS — LAWN & GARDEN
ALL DEALERS

	All Dealers	High Profit Dealers	Low Profit Dealers	Your Figures
Current Ratio	1.38 TO 1	1.21 TO 1	1.49 TO 1	
Ownership Equity	27.89 %	24.50 %	23.32 %	
Receivables Turnover	15.46 TIMES	9.55 TIMES	13.81 TIMES	
Net Working Capital Turnover	7.01 TIMES	7.58 TIMES	5.51 TIMES	
Inventory to Net Working Capital	2.98 TO 1	4.75 TO 1	2.59 TO 1	
Turnover of Total Assets	1.69 TIMES	1.14 TIMES	1.71 TIMES	
Return on Total Assets	5.87 %	11.46 %	(2.43 %)	
Return on Net Worth	21.05 %	46.77 %	(10.42 %)	

BALANCE SHEET AND INVENTORY STUDIES — LAWN & GARDEN
ALL DEALERS

BALANCE SHEET ACCOUNTS	AVERAGE ALL VOLUME GROUPS						HIGH PROFIT DEALERS			LOW PROFIT DEALERS		
Assets	Amount	% Total Assets	Turn Over	Amount	% Total Assets		Amount	% Total Assets	Turn Over	Amount	% Total Assets	Turn Over
Cash and Securities	$ 25,692	4.64					$ 25,437	3.73		$ 13,942	2.29	
Accounts and Notes Receivable	60,591	10.96					81,081	11.89		75,257	12.38	
TOTAL	86,283	15.60					106,518	15.62		89,199	14.67	
Inventories:												
New Equipment	270,413	48.90	1.49				370,247	54.33	.83	287,778	47.34	1.30
Used Equipment	40,508	7.33	2.40				28,059	4.11	2.04	46,737	7.69	2.31
Repair Parts	67,539	12.21	2.24				66,626	9.78	1.29	115,281	18.96	1.96
Other Lines	19,017	3.44	2.30				20,494	3.00	4.32	37,894	6.24	2.19
Rental or Lease	434	.08	8.57				114	.02	26.46	0	.00	.00
TOTAL INVENTORIES	397,911	71.96	1.76				485,540	71.24	1.12	487,690	80.23	1.62
TOTAL CURRENT ASSETS	484,194	87.56					592,058	86.86		576,889	94.90	
Fixed Assets	52,399	9.48					72,936	10.70		29,353	4.83	
Finance Reserve and Other Assets	16,352	2.96					16,580	2.44		1,626	.27	
TOTAL ASSETS	552,945	100.00					681,574	100.00		607,868	100.00	
Liabilities and Net Worth												
Current Accts. Payable & Accruals	94,549						96,109			75,218		
Floor Plan Notes Payable	224,737						355,120			285,677		
Notes Payable — Bank and Other	31,336						38,642			27,515		
Mortgages and Other Long Term	48,085						24,689			77,709		
TOTAL LIABILITIES	398,707						514,560			466,119		
Net Worth	154,238	27.89					167,014	24.50		141,749	23.32	
TOTAL LIABILITIES AND NET WORTH	552,945						681,574			607,868		

All averages herein are adjusted to nearest dollar, and percentages are adjusted to nearest 1/100th of a percent.

PERSONNEL STUDY — LAWN & GARDEN
ALL DEALERS

Number of Employees per Dealer (including Owners & Managers)	8.59
Total Sales per Employee	$109,029
Total Assets per Employee	$ 64,371
Number of Salesmen (including Owners & Managers)	2.22
New and Used Sales per Salesman	$261,192
Number of Partsmen	1.67
Parts Sales per Partsman	$131,023
Number of Servicemen	3.07
Service Sales per Serviceman	$ 23,701
Number of Office Employees	1.19
Number of Other Employees	.44

Financial records, carefully maintained, provide dealers with the only sound measuring device for business success. They help to measure the ability of their business to generate profits and to meet claims against it. They provide the basis, also, for isolating the business' strengths and weaknesses and may, through analysis, suggest how best to plan for profits and to maintain solvency.

With this in mind, the following ratios have been developed from year-end operating statements and balance sheets submitted by dealers in the current studies. Dealers having adequate records will find comparison of these ratios most revealing and helpful.

CURRENT RATIO

Figure by dividing Current Assets by Current Liabilities.

This ratio compares the total of Cash and the items which will be converted into cash, with debts which must be paid soon. Thus, the ratio reflects ability to handle current debts.

A ratio of 2.00 to 1 ($2.00 of Current Assets for each $1.00 of Current Liabilities) is considered satisfactory. When the ratio falls below 2.00 to 1, creditors have a greater interest in Current Assets than the owner.

EMPLOYEE STUDIES

In the blank column provided for your Region and Volume Group, enter the number of employees in each category as applicable for your business. Then — calculate the average Sales Per Employee by Dividing the applicable sales amount by the number of employees you have just entered for each category.

Comparison of your sales per employee in each category to the averages for your Region and Volume Group is the first step in measuring employee efficiency. The next step is to compare your current year figures to previous years for your business to identify improvements.

OWNERSHIP EQUITY

Figure by dividing Net Worth by Total Assets.

This ratio shows how much of the business is proprietor-owned; and conversely, by subtracting the percentage from 100%, it shows how much of the business is financed by outside capital (creditors).

When Net Worth is less than 50% of Total Assets, creditors have a greater interest in the assets than proprietors.

RECEIVABLES TURNOVER

Figure by dividing Total Sales by Receivables.

This ratio shows whether or not Receivables are in proportion to amount of business transacted. When lower than average, it might mean that collections are not up to par.

This ratio might be used every month to good advantage. Divide sales year-to-date by receivables and compare with ratio for same period in past two or three years. Unfavorable trends can thus be detected.

NET WORKING CAPITAL TURNOVER

Net Working Capital is determined by subtracting Current Liabilities from Current Assets.

Rate of turnover is figured by dividing Sales by Net Working Capital.

How many times Net Working Capital turns over in a year in Sales is a good measure of activity for any business. It is a sensitive indicator of overtrading or undertrading; and as such, it is a good test of the adequacy of Net Working Capital for the scale of operations of the firm.

When this ratio is too high, it will usually be found that the dealership is short of cash and heavy on receivables and paid-for inventory. This situation promotes borrowing or sale of equipment at a loss to meet payroll and other fixed obligations. When it is too low, inefficient use of capital in the dealership is indicated. It is usually attended by a pile-up of receivables or inventory or cash, or a combination of these, in excess of the needs of the business.

INVENTORY TO NET WORKING CAPITAL

Figure by dividing Inventory by Net Working Capital.

This ratio for most dealers in the farm and power equipment industry is extremely high because first, inventories are characteristically heavy and represent a major portion of Total Current Assets; and second, because of the heavy financing of whole goods.

In other retail lines where the ratio of Inventories to Net Working Capital is less than 100%, inventory requirements are relatively smaller and supplier credit terms are shorter.

TURNOVER OF TOTAL ASSETS

Figure by dividing Sales by Total Assets.

The importance of this ratio is illustrated in the question of why a dealership with $100,000 of Total Assets achieves Total Sales of $500,000 while another with an equal amount of assets achieves a Total Sales volume of only $300,000. The first would seem to be more efficient in its use of assets.

However, a high rate of turnover is not necessarily the key to profits. As in the ratio, Sales to Net Working Capital, high turnover may be an indicator of inadequate capital in relation to sales volume.

RETURN ON ASSETS

Figure by dividing Net Profit by Total Assets.

This ratio, too, is an important test of operating efficiency. It is a test of the quality of management in the utilization assets in the business. It lays aside all problems related to the source of funds, whether such funds are obtained from capital investment, borrowings, or purchases on credit. It shows the return on total capital invested in the business (owner capital plus outside capital), which is represented by Total Assets.

RETURN ON NET WORTH

Figure by dividing Net Profit by Net Worth.

In using this ratio, it is important that a fair salary for the owner be included in operating expenses of the business. The owner's time is certainly worth something; therefore, all Net Profit ratios should be based upon the profit *after* this salary is paid or taken into consideration in the figuring. Adequacy of profit must be measured in the light of current returns on money in the investment market, as well as compensation for risk.

INVENTORY TURNOVER

Figure by dividing cost of sales by Inventory.

This ratio shows the number of times Inventory is bought and sold during the year, reflected in "turnover" in the balance sheet studies.

Appendix B

Job Descriptions

The successful hiring of well-qualified employees can only proceed when the interviewer knows exactly what he is looking for. Following are job descriptions of the manpower required in a typical service department. These job descriptions are invaluable when performance evaluations are being made. Every organization should have a job description for all employees.

The following pages[1] describe the various jobs that can be found in many service departments. These descriptions were supplied by The John Deere Company.

---

[1] Courtesy of The John Deere Company

JOB DESCRIPTION: SERVICE MANAGER

JOB TITLE : SERVICE MANAGER

Department: Service Department

Reports To: General Manager of Dealership

Supervises: Shop Foreman, Mechanics, Set-Up and Pre-Delivery Men (usually 3-15 persons)

JOB FUNCTION:

Manages entire shop operations—schedules shop work, assigns locations and personnel to specific work orders, and maintains standards of efficiency and work output to satisfy customer service needs, predelivery service on new goods, and repair of trade-in equipment. Advises customers, determines the amount, and estimates the value of, shop work required for proper repair and makes recommendations to customers for operation and care of their equipment.

PRIMARY DUTIES:

1. Makes assignments in daily operations in shop functions, schedules priority of shop jobs, and plans for orderly flow of work. Plans total shop operations, and keeps detailed records of time and performance of employees.

2. Assigns mechanics to specific work orders through the Shop Foreman who supervises shop work.

3. Advises customers of necessary repair service including estimates of costs involved. Makes recommendations to customers for preventive maintenance requirements including proper care of equipment, and regular maintenance requirements.

4. Arranges for control of each shop job and the proper charge for labor and materials to the customer, or to the proper account for pre-delivery service or used-goods reconditioning.

5. Handles warranty claims including computation of charges for shop work. Handles customer complaints and orders rework, when justified, for proper repair.

6. Maintains shop tools and equipment in up-to-date and usable condition and maintains "good housekeeping" practices.

7. Assumes responsibility for prompt reconditioning of trade-in equipment.

8. Keeps himself up to date with all bulletins and service information received by the dealership and informs service personnel regarding such information as it concerns their individual assignments.

9. Maintains a library of service manuals, bulletins, and other service publications for Service Department use.

GENERAL:

The position of Service Manager requires organizational ability and systematic timing if it is to accomplish over-all efficiency in producing satisfactory shop work. It requires an adequate knowledge of work processes and requirements for both physical labor and shop equipment. The Service Manager must be capable of accurately judging time and cost involved in projected work assignments.

This important position requires leadership, tact and knowledge in assigning and instructing men in the proper performance of mechanical operations, safely, and without damage to materials or persons. In addition, the Service Manager should have a basic understanding of accounting systems and procedures so that he may properly prepare invoices, work orders, warranty claims, work schedules, time reports, and other miscellaneous records as required.

JOB DESCRIPTION: SHOP FOREMAN

JOB TITLE: SHOP FOREMAN

Department: Service Department

Reports To: Service Manager

Supervises: Work assignments of the entire shop staff as designated by Service Manager.

JOB FUNCTION:

Assigns and gives work direction to employees; participates in the repair and reconditioning of customers' machinery and used machines taken in trade, and maintains, operates and supervises shop equipment, tools, and delivery equipment.

PRIMARY DUTIES:

1. Assists in the appraisal of repair work coming into the shop and discusses with Service Manager the service required, both in parts and labor.

2. Assists in the direction and planning activity of the service employees in reconditioning and repairing machines; assigns jobs and instructs employees in their work.

3. Arranges for, and participates in, reconditioning of used equipment taken in trade.

4. Issues work orders containing pertinent job information and specifying job instructions, identifying customer and machine.

5. Checks shop jobs to make certain that work is properly and fully performed, and approves release of equipment for delivery to customer.

6. Checks mechanics' daily time sheets, verifying distribution of working time, and posts hours chargeable to customer jobs, used equipment items, and various in-shop functional duties. Instructs employees in the efficient and safe use of shop tools and equipment, and delivery equipment.

7. Instructs customers in proper field operation of machines purchased, and makes adjustments or repairs machines in the field.

GENERAL:

The Shop Foreman must be thoroughly familiar with John Deere products as well as similar products of competitors. He must know the detailed makeup of many machines in order to determine the repair necessary and must have the knowledge and skill to direct such repair.

He should have or acquire thorough knowledge of procedures and dealership policies, and have leadership ability to achieve the respect of shop employees and develop satisfactory cooperation.

JOB DESCRIPTION: MECHANIC

JOB TITLE: MECHANIC
(Serviceman) (Service Technician)

Department: Service Department

Reports To: Shop Foreman

Supervises: None

JOB FUNCTION:

Repairs and reconditions customers' machinery; reconditions used equipment for resale by the dealership. Maintains and operates shop machines and tools.

PRIMARY DUTIES:

1. Reconditions and repairs all engines, tractors and equipment assigned to him by the Shop Foreman. Diagnoses trouble, plans methods and sequence of performing work required. Uses a variety of shop tools and performs a certain amount of lifting.

2. Performs service on Diesel equipment, electrical systems, and hydraulic systems.

3. Keeps records required to establish cost of parts and service charges to the customer.

4. Handles customer problems both objectively and tactfully.

5. Maintains and cares for shop tools and equipment.

6. Assists in directing work of helpers or on-the-job trainees and provides instruction, as required.

7. Maintains a clean work area and performs miscellaneous duties associated with the job.

GENERAL:

The mechanic must be thoroughly familiar with tractors and equipment. His work is not supervised fully; therefore, he must know the detailed construction of many machines in order to determine the work required; he must have the knowledge and skill to make such repairs properly.

JOB DESCRIPTION: SET-UP PRE-DELIVERY MAN

JOB TITLE: SET-UP AND PREDELIVERY MAN

Department: Service Department

Reports To: Shop Foreman

Supervises: None

JOB FUNCTION:

Assembles and services all types of new tractors and equipment offered for sale by the dealership; prepares equipment for delivery to the customer.

PRIMARY DUTIES:

1. Assembles and services all types of new trac-tors, equipment, and other merchandise offered for sale by the dealership. Maintains necessary tools to perform such work.

2. Prepares all types of new and used equipment for sale or resale. This includes cleaning, paint-ing (by hand brushing or spraying) where necessary, of parts, assemblies, or complete machines.

3. Checks and reports malfunctions due to de-fective parts; records unusual noises, leaks, etc.

4. Lubricates and cleans equipment, maintains a clean work area, and performs other duties as-sociated with the job.

GENERAL:

He must have knowledge of tractors and equip-ment. His work is supervised, but the individual should be familiar with predelivery instructions and be able to follow both written and oral direc-tions in setting up equipment for delivery.

JOB TITLE: FIELD SERVICEMAN

Department: Service Department

Reports To: Service Manager

Supervises: None

JOB FUNCTION:

Repairs and reconditions customers' equipment at the customer's location—home or work area. Maintains and operates shop equipment and tools.

PRIMARY DUTIES:

1. Reconditions and repairs engines, tractors, and equipment at the customer's location, as assigned to him by the Service Manager. Diagnoses trouble, plans method and sequence of performing work required. Uses a variety of tools and is prepared to do some lifting.

2. Must be familiar with service requirements on Diesel equipment, electrical systems, and hydraulic systems.

3. Keeps required records to establish cost of parts, repair, and service charges to the customer.

4. Handles customer problems both objectively and tactfully.

5. Maintains and cares for shop tools and equipment, and service vehicles.

6. Operates trucks when necessary (holds chauffer's license if required).

GENERAL:

The Field Serviceman must be thoroughly familiar with tractors and equipment. His work is not generally supervised directly; therefore, he must know the detailed makeup of many machines in order to determine the repair work required. Must have the knowledge and skill to properly make such repairs as are necessary.

JOB DESCRIPTION: TRUCK DRIVER

JOB TITLE: TRUCK DRIVER

Department:　Service Department

Reports To:　Service Manager

Supervises:　None

JOB FUNCTION:

Operates truck to pick up and deliver tractors, equipment, machinery, parts and supplies to or from the dealership.

PRIMARY DUTIES:

1. Follows written or verbal instructions for the pickup, transport, and delivery of tractors, equipment, machinery, parts, and supplies to customers of the dealership. Properly loads and unloads the truck. Checks condition of equipment delivered to customers.

2. Gives basic operating instructions to the customer at time of delivery.

3. Operates trucks owned by the dealership (holds a chauffer's license).

4. Keeps records of pickups and deliveries; obtains necessary signatures and receipts, as required.

5. Keeps truck in good repair by performing necessary maintenance, such as changing tires, washing, adding fuel, changing oil, etc.

6. Performs other miscellaneous duties assigned to the job and to meet the needs of the Service Department.

GENERAL:

Individual must have excellent knowledge of trucks and be thoroughly familiar with their proper loading and operation. Since he operates the vehicles on public roads, he must be constantly alert in loading and in driving at all times. He must be capable of following directions and be courteous and helpful to customers. Should have sufficient knowledge to be able to give customers basic operating instructions when equipment is delivered.

Electronic Data Processing

The age of the computer is here. Businessman should acquire a basic understanding of what the computer is, what it isn't, and learn to use it intelligently. Some of the many ways this electronic marvel can assist you in operating your enterprise will be examined in the following pages.

The computer is made up of a processing unit along with a means of communicating with the processing unit and memory devices for storing information, displaying information, and manipulating stored data. It is capable of receiving data, storing data, recalling data, performing complex mathematical operations, and displaying or printing out the results in various formats. The electronic devices are referred to as hardware. Instructions to the unit may be given from a typewriter-like keyboard or from magnetic tapes, discs, or cassettes. These instructions, called programs, are referred to as software.

What is a computer? It is a combination of devices for making and displaying computations. It has the capability of doing these computations exceedingly fast and accurately. It is *not* a black magic box. It cannot do anything it is not told how to do. It knows nothing it has not been told. It is not a manager. Though it can record history, plot trends, and predict the future, the manager still has to implement decisions.

The computer may be "in-house"—that is, all components are located within the confines of the dealership—or it may be "remote," in which case all components are located away from the dealership. A combination where the receiving and sending unit are located in-house and the processor is located elsewhere is often used.

More important than the location of the computer is the availability of software. The problems of the dealership are specialized, requiring programs that are tailored to the retail equipment business. If the availability of programs and service are the same, the convenience and speed of the in-house computer cannot be disputed. The following are typical reports that can be generated by computer. The format and exact data cited will vary depending on the program.

Both hardware and software are available from several sources including private companies, your trade association, and your supplier. Shop for the one that best fits your needs. Strive to acquire a system that enables you to enter data only once yet allows it to flow through to many places it is needed in the accounting chain. The following printouts are shown to illustrate the wide range of functions that can be computerized.

The Balance Sheet, Operating Statement, Expense Statement, and Departmental Profit and Loss Statement are shown in Appendixes C.1–C.5. The expense allocation is shown by percentage. It could be individualized if that was considered necessary.

Appendix C.6 shows how funds are being used in the business. The changes in assets, liabilities, and net worth are indicators of the health of the business.

The good businessman budgets his activities. This gives him a road map and timetable to lead him on toward his goals. Work sheets for that purpose are shown in Appendix C.7. Other programs allow you to project number of units as well as dollars.

The computer keeps up with those figures and prints an Operating Trend and Budget Report. Variations from the budgetary amount point to areas of the business demanding attention. A typical report is shown in Appendixes C.8–C.10. One software program is also available which shows three-year operating trends. This is shown in Appendixes C.11–C.13.

Inventory control is always a problem. There are numerous programs available to keep up with whole goods and parts. The whole goods inventory sheet may take the form of a sales price book as shown in Appendix C.14. Available with this program is a wash-out sheet that traces the sale to a climax when all trade-ins are disposed. A typical washout sheet is shown in Appendix C.15. Other programs list the equipment with all extras added, including a place for assembly, handling, and freight charges (Appendix C.16). Vital to the cash flow position of the business is the aging of floor planned equipment. A typical report is shown in Appendix C.17.

The format for parts inventory control is quite varied. The locator system, the suggested stock order, and the exchange and movement analysis in Appendix C.18 are typical of the programs available for parts management. Parts sales can be written up by hand or entered directly into the computer terminal at the parts counter where an individual ticket will be printed. At the end of the billing period, a statement is prepared by the computer. A typical statement is shown in Appendix C.19.

The aged accounts receivable report shown in Appendix C.20 is a most important document. Current receivables are an asset the banks will lend on. A comparable document, the Aged Accounts Payable report, is also available as shown in Appendix C.21. Good credit is a definite asset.

Nowhere is employee productivity more difficult to keep up with than in the service shop. The reports shown in Appendix C.22 can be most helpful. The monthly productivity of the individual is based on flat rate pricing. Activities to date and, in addition, the skill level he has attained are given as well as a composite of all mechanics. Controllable and uncontrollable expenses for the shop are shown along with the labor recovery rate. Numerous other reports are also available.

Productivity of the partsmen and salesmen can be quantified in a similar manner.

Ratios are an integral part of financial management. Appendix C.23 shows a typical printout of four ratios. It shows the trends by the month with a comparison of the past year. The trend may be more important than the actual ratio.

A complete computer accounting software package will do all the ledgers normally prepared in a retail business. Since pay is such an important consideration of the employee, the employer, and the government, a typical Payroll and Earning report is shown in Appendix C.24.

The computer is capable of storing and retrieving data presented to it. Software is available to output the data in various formats. A word of caution is in order. Select the reports you need and use. Don't print out so many reports that you are intimidated by the volume of material you have to analyze. Finally, remember, the computer is a tool. You are still the manager.

The printouts that follow were furnished by the National Farm and Power Equipment Dealers Association, Dealer Information Systems, Corporation, and International Harvester Company. There are several other sources of good software.

```
DM03ST                    B A L A N C E   S H E E T              PAGE   1

DEALER 7-99-9999-9                              AS OF        MAY 30 1979
-------------------------------------------------------------------------------

    ACCT.                        A S S E T S          AMOUNT
    NO.


              CURRENT ASSETS

    101-00    PETTY CASH                               50.00
    102-00    CASH ON HAND                            200.00
    103-00    CASH IN BANK                          1,232.20
              TOTAL CASH                            1,482.20  *

              ACCTS RECEIVABLES

    104-00    ACCTS REC-PARTS                      15,351.55
    104-01    ACCTS REC-MACHINE                    15,586.70
    105-01    CONTRACTS IN TRANSIT                 20,000.00
    106-00    DEALER RESERVE-CASE                   1,365.19
    106-01    DEALER RESERVE-RNTL                   3,313.87
    106-02    DEALER RESERVE-FIN CHGS               3,452.10
    106-03    DEALER RESERVE-UNEARNED               3,452.10
    108-00    FED INC TAX REFUND                    1,465.48
              NET RECEIVABLES                      63,986.99  *

              INVENTORIES

    500-00    NEW TRACTORS                        692,711.90
    501-00    NEW IMPLEMENTS                      116,000.00
    502-00    GARDEN EQUIPMENT                           .00
    504-00    NEW PARTS                           130,520.33
    506-00    ACCESSORIES                           5,797.00
    514-00    USED EQUIP                          277,745.99
              TOTAL INVENTORIES                 1,222,775.22  *

              PREPAID ITEMS

    153-00    PREPAID PPTY TAX                        341.25
    154-00    DEPOSITS-WRKMS COMP                     475.00
    156-00    CORP INC TAX DEPOSIT                    935.00
              TOTAL PREPAID ITEMS                   1,751.25  *

              TOTAL CURRENT ASSETS              1,289,995.66  **

              FIXED ASSETS

    111-00    LEASEHOLD IMP                           498.50
```

DM03ST   B A L A N C E   S H E E T   PAGE   2

DEALER 7-99-9999-9   AS OF   MAY 30 1979

ACCT.   A S S E T S   AMOUNT
  NO.

111-01   ALLW FOR DEPRN   302.00-
112-00   FURNITURE & FIXTURES   521.04
112-01   ALLW FOR DEPRN   314.00-
113-00   SHOP EQUIP & TOOLS   4,080.56
113-01   ALLW FOR DEPRN   1,620.00-
114-00   OFFICE EQUIP   1,714.68
114-01   ALLW FOR DEPRN   874.00-
115-00   AUTO & TRUCKS   4,112.00
115-01   ALLW FOR DEPRN   3,979.25-
116-00   RENTAL EQUIP   50,000.00
116-01   ALLW FOR DEPRN   10,000.00-
            NET FIXED ASSETS   43,837.53 *

            TOTAL ASSETS   1,333,833.19 ***

```
DM03ST                    B A L A N C E   S H E E T              PAGE    3

DEALER 7-99-9999-9                            AS OF          MAY 30 1979
```

```
ACCT.                    L I A B I L I T I E S           AMOUNT
 NO.


           CURRENT LIABILITIES

201-00     ACCTS PAY-OTHER                    237,150.08
201-01     ACCTS PAY-CASE                     300,245.00
202-00     NOTES PAY-CASE                     169,620.34
202-01     NOTES PAY-ALLIED                    22,848.62
202-02     NOTES PAY-OTHER                           .00
202-03     NOTES PAYABLE BANK                   2,000.00
202-04     NTS PAYABLE-AT PETERSON             7,500.00
203-00     SALES TAX PAY                        2,131.70
207-00     WITHHOLDING-FED                      4,291.17
207-01     WITHHOLDING-STATE                      492.12
208-00     ACCRUED SALARIES                     3,000.00
208-01     ACCRUED VACATION PAY                 1,000.00
208-02     ACCRUED INC PAYABLE                    569.07
211-00     FEDERAL INC TAX PAYABLE              1,200.00
212-00     CALIF FRAN TAX PAY                      18.00
215-00     FICA TAX PAY                           561.85
216-00     UNEMPL TAX PAY                          84.82
217-00     STATE DISABL PAY                         1.38
218-00     DEPOSIT ON EQUIP                     1,000.00
219-00     DIVIDEND PAYABLE                     5,510.00
           TOTAL CURRENT LIABIL               759,224.15  *

           LONG TERM LIABILITIES

240-00     LONG TERM N/P                             .00
241-00     MORTGAGES PAYABLE                  145,000.00
           TOTAL LONG TERM                    145,000.00  *

           TOTAL LIABILITIES                  904,224.15  **

           NET WORTH

261-00     CAPITAL STOCK                       32,450.00
262-00     PAID IN SURPLUS                        643.94
263-00     RETAINED EARNINGS                  162,569.64
299-99     PRIOR YEAR PROFIT/LOSS                    .00
           CURRENT YEAR P & L                 233,945.46

           TOTAL NET WORTH                    429,609.04  *

       TOTAL LIABILITIES & NET WORTH        1,333,833.19  ***
```

DM03ST
EXPENSE STATEMENT
PAGE 1

DEALER 7-99-9999-9   5-MONTH PERIOD   AS OF   MAY 30 1979

ACCT. NO.		CURRENT MONTH	%	YEAR-TO-DATE	%
	EXPENSES				
31-00	SALARIES-MGRS	1,383.54	.71	8,700.02	.68
32-00	SALARIES-OFFICE EMPL	752.70	.38	3,763.50	.30
33-00	SALARIES-SALESMEN	329.66	.17	2,730.96	.21
34-00	SALARIES-PARTSMAN	3,102.40	1.58	16,594.66	1.30
35-00	ELECTRONIC ACCTG SER	.00	.00-	517.38	.04
36-00	RENT & LEASE EXP	1,000.00	.51	6,478.38	.51
36-01	LEASE EQUIP	.00	.00-	.00	.00-
37-00	UTILITIES	500.00	.26	2,806.24	.22
38-00	TELEPHONE	200.00	.10	1,000.00	.08
39-00	OFFICE & STORE SUPPLY	50.00	.03	593.76	.05
40-00	EMPLOYEE BENEFITS	160.00	.08	900.00	.06
41-00	FREIGHT MISC	50.00	.03	651.42	.05
42-00	SHOP SUPPLY & EQ REP	25.00	.01	622.58	.05
43-00	CAR & TRUCK EXP	25.00	.01	144.00	.01
43-01	GAS & OIL	10.00	.01	264.32	.02
45-00	TRAVEL & SALES	50.00	.03	250.00	.02
46-00	ADVERTISING	500.00	.26	3,161.00	.25
47-00	PAYROLL-TAXES	.00	.00-	594.38	.05
49-00	TAXES-BUS PPTY	.00	.00-	.00	.00-
51-00	OTHER TAX & LIC	.00	.00-	.00	.00-
52-00	DUES & SUBSCRIPT	.00	.00-	.00	.00-
53-00	INSURANCE	1,000.00	.51	5,635.80	.44
54-00	LEGAL & AUDITING	.00	.00-	827.32	.06
55-00	WORKMENS COMP INSUR	50.00	.03	250.00	.02
56-00	BANK CHARGES	.00	.00-	.28	.00-
57-00	GROUP INSURANCE	.00	.00-	.00	.00-
59-00	REPAIRS & MAINT-BLDG	100.00	.05	924.52	.07
60-00	DEPRECIATION	.00	.00-	.00	.00-
61-00	BAD DEBTS	.00	.00-	.00	.00-
62-00	BAD DEBTS COLLECTION C	.00	.00-	.00	.00-
64-00	INTEREST	.00	.00-	560.44	.04
67-00	DONATIONS	.00	.00-	.00	.00-
68-00	AFTERSALE WARR-NEW EQ	1,000.00	.51	5,000.00	.39
69-00	AFTERSALE WARR-USED	500.00	.26	2,500.00	.20
70-00	AFTERSALE ADJ-SERVICE	1,000.00	.51	5,000.00	.39
71-00	MISCELLANEOUS EXP	500.00	.26	2,510.12	.20
	TOTAL EXPENSE	12,238.30	6.27	72,882.08	5.72 ***

ACCOUNT NO. SALES COST	ACCOUNT	CURRENT MONTH SALES	COST	MARGIN	%	YEAR SALES	TO COST	DATE MARGIN	%
				NEW EQUIPMENT DEPARTMENT					
300-00 400-00	NEW TRACTORS	60000.00	48000.00	12000.00	20.00	405680.94	332465.72	73215.22	18.04
301-00 401-00	NEW IMPLEMENTS	20000.00	16000.00	4000.00	20.00	145760.82	117488.37	28272.45	19.39
	TOTAL NEW EQUIPMENT	80000.00	64000.00	16000.00	20.00	551441.76	449954.09	101487.67	18.40
				USED EQUIPMENT DEPARTMENT					
314-00 414-00	USED TRACTORS	40000.00	30000.00	10000.00	25.00	250967.72	195832.91	55134.81	21.96
315-00 415-00	USED IMPLEMENTS	1000.00	500.00	500.00	50.00	5000.00	2500.00	2500.00	50.00
	TOTAL USED EQUIPMENT	41000.00	30500.00	10500.00	25.60	255967.72	198332.91	57634.81	22.51
	TOTAL NEW AND USED	121000.00	94500.00	26500.00	21.90	807409.48	648287.00	159122.48	19.70
				PARTS DEPARTMENT					
304-00 404-00	PARTS-CUSTOMER	20000.00	15000.00	5000.00	25.00	147847.93	116155.70	31692.23	21.43
305-00 405-00	PARTS-INTERNAL	5000.00	5000.00	.00	.00	55928.77	46826.57	9102.20	16.27
	TOTAL PARTS	25000.00	20000.00	5000.00	20.00	203776.70	162982.27	40794.43	20.01
				SERVICE DEPARTMENT					
318-00 418-00	SERVICE LAB-CUST	10000.00	3352.74	6647.26	66.47	60200.00	31341.30	28858.70	47.93
319-00	SERVICE LAB-INTERNAL	10000.00	.00	10000.00	100.00	51752.00	.00	51752.00	100.00
	TOTAL SERVICE	20000.00	3352.74	16647.26	83.23	111952.00	31341.30	80610.70	72.00
				RENTAL DEPT.					
331-00 431-00	RENTAL-NON-TAXABLE	30000.00	25500.00	4500.00	15.00	151950.00	127500.00	24450.00	16.09
	TOTAL RENTAL DEPT	30000.00	25500.00	4500.00	15.00	151950.00	127500.00	24450.00	16.09
				ALL DEPARTMENTS					
	GRAND TOTAL	196000.00	143352.74	52647.26	26.86	1275088.18	970110.57	304977.61	23.91
	LESS TOTAL EXPENSE			12288.30				72882.08	
	OPERATING PROFIT OR LOSS			40358.96	20.59			232095.53	18.20
				OTHER INCOME & DEDUCTIONS					
351-00	INTEREST EARNED			304.44	.15			1699.16	.13
352-00	CASH DISCTS EARNED			.00	.00			150.77	.01
353-00	FINANCE CHGS EARNED			.00	.00			.00	.00
380-00	SUNDRY GAINS/LOSSES			.00	.00			.00	.00
382-00	BAD DEBTS RECOVERED			.00	.00			.00	.00
384-00	GAIN ON FIXED ASSET			.00	.00			.00	.00
385-00	CASH OVER & SHORT			.00	.00			.00	.00
386-00	OTHER COMMS & EARN			.00	.00			.00	.00
387-00	INCOME CASE PROGRAM			.00	.00			.00	.00
	TOTAL OTHER INCOME-DEDUC			304.44	.15			1849.93	.14
	TOTAL NET PROFIT OR LOSS			40663.40	20.74			233945.46	18.34

DM06ST   MONTHLY   DEPARTMENTAL PROFIT & LOSS STATEMENT   PAGE   1

DEALER 7-99-9999-9   MONTH ENDING   MAY 30 1979

                                       MONTH ENDING   MAY 30 1979

DESCRIPTION	ALLOCATION PERCENTAGES	MONTH TOTAL	%	WHOLEGOODS DEPT.		PARTS DEPT.		SERVICE DEPT.		RENTAL DEPT.			
SALES		196000.00		121000.00		25000.00		20000.00		30000.00		.00	
COST OF SALES		143352.74	73.1	94500.00	78.0	20000.00	80.0	3352.74	16.7	25500.00	85.0	.00	.0
OPERATING MARGIN		52647.26	26.8	26500.00	21.9	5000.00	20.0	16647.26	83.2	4500.00	15.0	.00	.0
SALARIES-MGRS	25-25-25-25-00	1383.54	.7	345.88	.2	345.88	1.3	345.88	1.7	345.88	1.1	.00	.0
SALARIES-OFFICE EMPL	50-15-15-20-00	752.70	.3	376.35	.3	112.90	.4	112.90	.5	150.54	.5	.00	.0
SALARIES-SALESMEN	25-25-25-25-00	329.66	.1	82.41	.0	82.41	.3	82.41	.4	82.41	.2	.00	.0
SALARIES-PARTSMAN	00-100-0-00-00	3102.40	1.5	.00	.0	3102.40	12.4	.00	.0	.00	.0	.00	.0
ELECTRONIC ACCTG SER	25-25-25-25-00	.00	.0	.00	.0	.00	.0	.00	.0	.00	.0	.00	.0
RENT & LEASE EXP	00-00-00-100-0	1000.00	.5	.00	.0	.00	.0	.00	.0	1000.00	3.3	.00	.0
LEASE EQUIP	00-00-00-100-0	.00	.0	.00	.0	.00	.0	.00	.0	.00	.0	.00	.0
UTILITIES	25-25-25-25-00	500.00	.2	125.00	.1	125.00	.5	125.00	.6	125.00	.4	.00	.0
TELEPHONE	25-25-25-25-00	200.00	.1	50.00	.0	50.00	.2	50.00	.2	50.00	.1	.00	.0
OFFICE & STORE SUPPLY	25-25-25-25-00	50.00	.0	12.50	.0	12.50	.0	12.50	.0	12.50	.0	.00	.0
EMPLOYEE BENEFITS	25-25-25-25-00	160.00	.0	40.00	.0	40.00	.1	40.00	.2	40.00	.1	.00	.0
FREIGHT MISC	75-10-10-05-00	50.00	.0	37.50	.0	5.00	.0	5.00	.0	2.50	.0	.00	.0
SHOP SUPPLY & EQ RFP	10-35-30-25-00	25.00	.0	2.50	.0	8.75	.0	7.50	.0	6.25	.0	.00	.0
CAR & TRUCK EXP	25-25-25-25-00	25.00	.0	6.25	.0	6.25	.0	6.25	.0	6.25	.0	.00	.0
GAS & OIL	25-25-25-25-00	10.00	.0	2.50	.0	2.50	.0	2.50	.0	2.50	.0	.00	.0
TRAVEL & SALES	00-00-00-00-00	50.00	.0	.00	.0	.00	.0	.00	.0	.00	.0	.00	.0
ADVERTISING	33-00-33-34-00	500.00	.2	165.00	.1	.00	.0	165.00	.8	170.00	.5	.00	.0
PAYROLL-TAXES	25-25-25-25-00	.00	.0	.00	.0	.00	.0	.00	.0	.00	.0	.00	.0
TAXES-BUS PPTY	25-25-25-25-00	.00	.0	.00	.0	.00	.0	.00	.0	.00	.0	.00	.0
OTHER TAX & LIC	25-25-25-25-00	.00	.0	.00	.0	.00	.0	.00	.0	.00	.0	.00	.0
DUES & SUBSCRIPT	25-25-25-25-00	.00	.0	.00	.0	.00	.0	.00	.0	.00	.0	.00	.0
INSURANCE	25-25-25-25-00	1000.00	.5	250.00	.2	250.00	1.0	250.00	1.2	250.00	.8	.00	.0
LEGAL & AUDITING	25-25-25-25-00	.00	.0	.00	.0	.00	.0	.00	.0	.00	.0	.00	.0
WORKMENS COMP INSUR	25-25-25-25-00	50.00	.0	12.50	.0	12.50	.0	12.50	.0	12.50	.0	.00	.0
BANK CHARGES	25-25-25-25-00	.00	.0	.00	.0	.00	.0	.00	.0	.00	.0	.00	.0
GROUP INSURANCE	25-25-25-25-00	.00	.0	.00	.0	.00	.0	.00	.0	.00	.0	.00	.0
REPAIRS & MAINT-BLDG	100-0-00-00-00	100.00	.0	100.00	.0	.00	.0	.00	.0	.00	.0	.00	.0
DEPRECIATION	50-00-00-50-00	.00	.0	.00	.0	.00	.0	.00	.0	.00	.0	.00	.0
BAD DEBTS	33-33-34-00-00	.00	.0	.00	.0	.00	.0	.00	.0	.00	.0	.00	.0
BAD DEBTS COLLECTION C	33-33-34-00-00	.00	.0	.00	.0	.00	.0	.00	.0	.00	.0	.00	.0
INTEREST	25-25-25-25-00	.00	.0	.00	.0	.00	.0	.00	.0	.00	.0	.00	.0
DONATIONS	40-20-20-20-00	.00	.0	.00	.0	.00	.0	.00	.0	.00	.0	.00	.0
AFTERSALE WARR-NEW EQ	100-0-00-00-00	1000.00	.5	1000.00	.8	.00	.0	.00	.0	.00	.0	.00	.0
AFTERSALE WARR-USED	100-0-00-00-00	500.00	.2	500.00	.4	.00	.0	.00	.0	.00	.0	.00	.0
AFTERSALE ADJ-SERVICE	00-00-100-0-00	1000.00	.5	.00	.0	.00	.0	1000.00	5.0	.00	.0	.00	.0
MISCELLANEOUS EXP	25-25-25-25-00	500.00	.2	125.00	.1	125.00	.5	125.00	.6	125.00	.4	.00	.0
TOTAL OPERATING EXPENSE		12288.30	6.2	3233.39	2.6	4281.09	17.1	2342.44	11.7	2381.33	7.9	.00	.0
NET OPERATING PROFIT/LOSS		40358.96		23266.61		718.91		14304.82		2118.67		.00	

OTHER INCOME-DEDUCTIONS   304.44
TOTAL NET PROFIT/LOSS   40663.40

DESCRIPTION	ALLOCATION PERCENTAGES	Y-T-D TOTAL	%	WHOLEGOODS DEPT.	%	PARTS DEPT.	%	SERVICE DEPT.	%	RENTAL DEPT.	%		%
SALES		1275088.18		807409.48		203776.70		111952.00		151950.00		.00	
COST OF SALES		970110.57	76.0	648287.00	80.2	162982.27	79.9	31341.30	27.9	127500.00	83.9	.00	.0
OPERATING MARGIN		304977.61	23.9	159122.48	19.7	40794.43	20.0	80610.70	72.0	24450.00	16.0	.00	.0
SALARIES-MGRS	25-25-25-25-00	8700.02	.6	2175.00	.2	2175.00	1.0	2175.00	1.9	2175.00	1.4	.00	.0
SALARIES-OFFICE EMPL	50-15-15-20-00	3763.50	.2	1881.75	.2	564.52	.2	564.52	.5	752.70	.4	.00	.0
SALARIES-SALESMEN	25-25-25-25-00	2730.96	.2	682.74	.0	682.74	.3	682.74	.6	682.74	.4	.00	.0
SALARIES-PARTSMAN	00-100-0-00-00	16594.66	1.3	.00	.0	16594.66	8.1	.00	.0	.00	.0	.00	.0
ELECTRONIC ACCTG SER	25-25-25-25-00	517.38	.0	129.34	.0	129.34	.0	129.34	.1	129.34	.0	.00	.0
RENT & LEASE EXP	00-00-00-100-0	6478.38	.5	.00	.0	.00	.0	.00	.0	6478.38	4.2	.00	.0
LEASE EQUIP	00-00-00-100-0	.00	.0	.00	.0	.00	.0	.00	.0	.00	.0	.00	.0
UTILITIES	25-25-25-25-00	2806.24	.2	701.56	.0	701.56	.3	701.56	.6	701.56	.4	.00	.0
TELEPHONE	25-25-25-25-00	1000.00	.0	250.00	.0	250.00	.1	250.00	.2	250.00	.1	.00	.0
OFFICE & STORE SUPPLY	25-25-25-25-00	593.76	.0	148.44	.0	148.44	.0	148.44	.1	148.44	.0	.00	.0
EMPLOYEE BENEFITS	25-25-25-25-00	800.00	.0	200.00	.0	200.00	.0	200.00	.1	200.00	.1	.00	.0
FREIGHT MISC	75-10-10-05-00	651.42	.0	488.56	.0	65.14	.0	65.14	.0	32.57	.0	.00	.0
SHOP SUPPLY & EQ REP	10-35-30-25-00	623.58	.0	62.35	.0	218.25	.1	187.07	.1	155.89	.1	.00	.0
CAR & TRUCK EXP	25-25-25-25-00	144.00	.0	36.00	.0	36.00	.0	36.00	.0	36.00	.0	.00	.0
GAS & OIL	25-25-25-25-00	264.32	.0	66.08	.0	66.08	.0	66.08	.0	66.08	.0	.00	.0
TRAVEL & SALES	00-00-00-00-00	250.00	.0	.00	.0	.00	.0	.00	.0	.00	.0	.00	.0
ADVERTISING	33-00-33-34-00	3161.00	.2	1043.13	.1	.00	.0	1043.13	.9	1074.74	.7	.00	.0
PAYROLL-TAXES	25-25-25-25-00	594.38	.0	148.59	.0	148.59	.0	148.59	.1	148.59	.0	.00	.0
TAXES-BUS PPTY	25-25-25-25-00	.00	.0	.00	.0	.00	.0	.00	.0	.00	.0	.00	.0
OTHER TAX & LIC	25-25-25-25-00	.00	.0	.00	.0	.00	.0	.00	.0	.00	.0	.00	.0
DUES & SUBSCRIPT	25-25-25-25-00	.00	.0	.00	.0	.00	.0	.00	.0	.00	.0	.00	.0
INSURANCE	25-25-25-25-00	5635.80	.4	1408.95	.1	1408.95	.6	1408.95	1.2	1408.95	.9	.00	.0
LEGAL & AUDITING	25-25-25-25-00	827.32	.0	206.83	.0	206.83	.1	206.83	.1	206.83	.1	.00	.0
WORKMENS COMP INSUR	25-25-25-25-00	250.00	.0	62.50	.0	62.50	.0	62.50	.0	62.50	.0	.00	.0
BANK CHARGES	25-25-25-25-00	.28	.0	.07	.0	.07	.0	.07	.0	.07	.0	.00	.0
GROUP INSURANCE	25-25-25-25-00	.00	.0	.00	.0	.00	.0	.00	.0	.00	.0	.00	.0
REPAIRS & MAINT-BLDG	100-0-00-00-00	924.52	.0	924.52	.1	.00	.0	.00	.0	.00	.0	.00	.0
DEPRECIATION	50-00-00-50-00	.00	.0	.00	.0	.00	.0	.00	.0	.00	.0	.00	.0
BAD DEBTS	33-33-34-00-00	.00	.0	.00	.0	.00	.0	.00	.0	.00	.0	.00	.0
BAD DEBTS COLLECTION C	33-33-34-00-00	.00	.0	.00	.0	.00	.0	.00	.0	.00	.0	.00	.0
INTEREST	25-25-25-25-00	560.44	.0	140.11	.0	140.11	.0	140.11	.1	140.11	.0	.00	.0
DONATIONS	40-20-20-20-00	.00	.0	.00	.0	.00	.0	.00	.0	.00	.0	.00	.0
AFTERSALE WARR-NEW EQ	100-0-00-00-00	5000.00	.3	5000.00	.6	.00	.0	.00	.0	.00	.0	.00	.0
AFTERSALE WARR-USED	100-0-00-00-00	2500.00	.1	2500.00	.3	.00	.0	.00	.0	.00	.0	.00	.0
AFTERSALE ADJ-SERVICE	00-00-100-0-00	5000.00	.3	.00	.0	.00	.0	5000.00	4.4	.00	.0	.00	.0
MISCELLANEOUS EXP	25-25-25-25-00	2510.12	.1	627.53	.0	627.53	.3	627.53	.5	627.53	.4	.00	.0
TOTAL OPERATING EXPENSE		72882.08	5.7	18884.05	2.3	24426.31	11.9	13843.60	12.3	15478.02	10.1	.00	.0
NET OPERATING PROFIT/LOSS		232095.53		140238.43		16368.12		66767.10		8971.98		.00	
OTHER INCOME-DEDUCTIONS		1849.93											
TOTAL NET PROFIT/LOSS		233945.46											

```
DM07ST                          APPLICATION OF FUNDS                    PAGE NO.  1
DEALER 7-99-9999-9                BALANCE SHEET
                               AS OF       MAY 30 1979
-----------------------------------------------------------------------------------------------

A S S E T S
                                 BEGINNING BALANCE          CURRENT MONTH              INCREASE
ACCT NO      DESCRIPTION        * CURRENT FISCAL YEAR *    * YTD BALANCE     *        * DECREASE- *


              CURRENT ASSETS
101-00 PETTY CASH                      50.00                      50.00
102-00 CASH ON HAND                   200.00                     200.00
103-00 CASH IN BANK                 1,500.00                   1,232.20
         TOTAL CASH              *  $1,750.00                          $1,482.20         $267.80-

              ACCTS RECEIVABLES
104-00 ACCTS REC-PARTS             18,661.20                  15,351.55
104-01 ACCTS REC-MACHINE           15,586.70                  15,586.70
105-01 CONTRACTS IN TRANSIT             .00                   20,000.00
106-00 DEALER RESERVE-CASE          1,365.19                   1,365.19
106-01 DEALER RESERVE-RNTL          3,313.87                   3,313.87
106-02 DEALER RESERVE-FIN CHGS      3,452.10                   3,452.10
106-03 DEALER RESERVE-UNEARNED      3,452.10                   3,452.10
108-00 FED INC TAX REFUND           1,465.48                   1,465.48
         NET RECEIVABLES         * $47,296.64                         $63,986.99      $16,690.35

              INVENTORIES
500-00 NEW TRACTORS               175,211.90                 692,711.90
501-00 NEW IMPLEMENTS              40,000.00                 116,000.00
502-00 GARDEN EQUIPMENT                 .00                        .00
504-00 NEW PARTS                   36,520.33                 130,520.33
506-00 ACCESSORIES                    797.00                   5,797.00
514-00 USED EQUIP                 100,819.40                 277,745.99
         TOTAL INVENTORIES       * $353,348.63                      $1222,775.22     $869,426.59

              PREPAID ITEMS
153-00 PREPAID PPTY TAX               341.25                     341.25
154-00 DEPOSITS-WRKMS COMP            475.00                     475.00
156-00 CORP INC TAX DEPOSIT           935.00                     935.00
         TOTAL PREPAID ITEMS     *  $1,751.25                          $1,751.25          $.00

         TOTAL CURRENT ASSETS    ** $404,146.52                     $1289,995.66     $885,849.14

              FIXED ASSETS
111-00 LEASEHOLD IMP                  498.50                     498.50
111-01 ALLW FOR DEPRN                 302.00-                    302.00-
112-00 FURNITURE & FIXTURES           521.04                     521.04
112-01 ALLW FOR DEPRN                 314.00-                    314.00-
113-00 SHOP EQUIP & TOOLS           4,080.56                   4,080.56
113-01 ALLW FOR DEPRN               1,620.00-                  1,620.00-
114-00 OFFICE EQUIP                 1,714.68                   1,714.68
114-01 ALLW FOR DEPRN                 874.00-                    874.00-
115-00 AUTO & TRUCKS                4,112.00                   4,112.00
115-01 ALLW FOR DEPRN               3,979.25-                  3,979.25-
116-00 RENTAL EQUIP                     .00                   50,000.00
116-01 ALLW FOR DEPRN                   .00                   10,000.00-
```

```
DM07ST                         APPLICATION OF FUNDS                    PAGE NO.   2
DEALER 7-99-9999-9                 BALANCE SHEET
                               AS OF        MAY 30 1979

--------------------------------------------------------------------------------------

A S S E T S
                               BEGINNING BALANCE         CURRENT MONTH             INCREASE
ACCT NO      DESCRIPTION     * CURRENT FISCAL YEAR *    * YTD BALANCE      *      * DECREASE- *

       NET FIXED ASSETS             *     $3,837.53           $43,837.53         $40,000.00

       TOTAL ASSETS              ****   $407,984.05         $1333,833.19         $925,849.14
```

L I A B I L I T I E S   A N D   N E T   W O R T H

ACCT NO	DESCRIPTION	BEGINNING BALANCE * CURRENT FISCAL YEAR *		CURRENT MONTH * YTD BALANCE *		INCREASE * DECREASE- *
	CURRENT LIABILITIES					
201-00	ACCTS PAY-OTHER	2,150.08		237,150.08		
201-01	ACCTS PAY-CASE	245.00		300,245.00		
202-00	NOTES PAY-CASE	169,620.34		169,620.34		
202-01	NOTES PAY-ALLIED	22,848.62		22,848.62		
202-02	NOTES PAY-OTHER	.00		.00		
202-03	NOTES PAYABLE BANK	2,000.00		2,000.00		
202-04	NTS PAYABLE-AT PETERSON	7,500.00		7,500.00		
203-00	SALES TAX PAY	1,326.70		2,131.70		
207-00	WITHHOLDING-FED	636.40		4,291.17		
207-01	WITHHOLDING-STATE	106.20		492.12		
208-00	ACCRUED SALARIES	.00		3,000.00		
208-01	ACCRUED VACATION PAY	.00		1,000.00		
208-02	ACCRUED INC PAYABLE	569.07		569.07		
211-00	FEDERAL INC TAX PAYABLE	.00		1,200.00		
212-00	CALIF FRAN TAX PAY	482.00-		18.00		
215-00	FICA TAX PAY	610.64-		561.85		
216-00	UNEMPL TAX PAY	315.18-		84.82		
217-00	STATE DISABL PAY	1.62-		1.38		
218-00	DEPOSIT ON EQUIP	.00		1,000.00		
219-00	DIVIDEND PAYABLE	6,490.00-		5,510.00		
	TOTAL CURRENT LIABIL	*	$199,102.97		$759,224.15	$560,121.18
	LONG TERM LIABILITIES					
240-00	LONG TERM N/P	.00		.00		
241-00	MORTGAGES PAYABLE	.00		145,000.00		
	TOTAL LONG TERM	*	$.00		$145,000.00	$145,000.00
	TOTAL LIABILITIES	**	$199,102.97		$904,224.15	$705,121.18
	NET WORTH					
261-00	CAPITAL STOCK	32,450.00		32,450.00		
262-00	PAID IN SURPLUS	643.94		643.94		
263-00	RETAINED EARNINGS	162,569.64		162,569.64		
299-99	PRIOR YEAR PROFIT/LOSS	.00		.00		
	CURRENT YEAR P & L	13,217.50		233,945.46		
	TOTAL NET WORTH	**	$208,881.08		$429,609.04	$220,727.96
	TOTAL LIABILITIES & NET WORTH	**	$407,984.05		$1333,833.19	$925,849.14

BUDGET PROJECTION WORKSHEET
FISCAL YEAR ENDING DECEMBER, 1979

ACCOUNT NO. SALES	COST	ACCOUNT	...LAST YEAR... SALES	MARGIN	%	...YOUR PROJECTION FOR CURRENT YEAR... SALES	COST	MARGIN/EXPENSE	%	
300-00	400-00	NEW TRACTORS	345,680	61,215	17.71	$............	$............	$............	......	300202
301-00	401-00	NEW IMPLEMENTS	125,760	24,272	19.30	$............	$............	$............	......	300302
314-00	414-00	USED TRACTORS	210,967	45,135	21.39	$............	$............	$............	......	300802
315-00	415-00	USED IMPLEMENTS	4,000	2,000	50.00	$............	$............	$............	......	300902
304-00	404-00	PARTS-CUSTOMER	127,847	26,692	20.88	$............	$............	$............	......	301502
305-00	405-00	PARTS-INTERNAL	50,928	9,102	17.87	$............	$............	$............	......	301602
318-00	418-00	SERVICE LAB-CUST	50,200	36,790	73.29	$............	$............	$............	......	302102
319-00	–	SERVICE LAB-INTERNAL	41,752	0	0.00	$............	$............	$............	......	302201
331-00	431-00	RENTAL-NON-TAXABLE	121,950	19,950	16.36	$............	$............	$............	......	302702
*TOTAL SALES-COST-MARGINS			* 1079,084	* 225,156	20.87	....	....	....	......	
	031-00	SALARIES-MGRS		7,314-	0.68			$............	......	400100
	032-00	SALARIES-OFFICE EMPL		3,008-	0.28			$............	......	400200
	033-00	SALARIES-SALESMEN		2,398-	0.22			$............	......	400300
	034-00	SALARIES-PARTSMAN		13,490-	1.25			$............	......	400400
	035-00	ELECTRONIC ACCTG SER		517-	0.05			$............	......	400500
	036-00	RENT & LEASE EXP		5,478-	0.51			$............	......	400600
	036-01	LEASE EQUIP		0	0.00			$............	......	400700
	037-00	UTILITIES		2,306-	0.21			$............	......	400800
	038-00	TELEPHONE		800-	0.07			$............	......	400900
	039-00	OFFICE & STORE SUPPLY		543-	0.05			$............	......	401000
	040-00	EMPLOYEE BENEFITS		640-	0.06			$............	......	401100
	041-00	FREIGHT MISC		601-	0.06			$............	......	401200
	042-00	SHOP SUPPLY & EQ REP		598-	0.06			$............	......	401300
	043-00	CAR & TRUCK EXP		119-	0.01			$............	......	401400
	043-01	GAS & OIL		254-	0.02			$............	......	401500

DM20ST
DEALER 7-99-9999-9

BUDGET PROJECTION WORKSHEET
FISCAL YEAR ENDING DECEMBER, 1979

ACCOUNT NO. SALES COST	ACCOUNT	...LAST YEAR... SALES	MARGIN	%	...YOUR PROJECTION FOR CURRENT YEAR... SALES COST	MARGIN/EXPENSE %	
045-00	TRAVEL & SALES		200-	0.02	$.............	....	401600
046-00	ADVERTISING		2,661-	0.25	$.............	....	401700
047-00	PAYROLL-TAXES		594-	0.06	$.............	....	401800
049-00	TAXES-BLS PPTY		0	0.00	$.............	....	401900
051-00	OTHER TAX & LIC		0	0.00	$.............	....	402000
052-00	DUES & SUBSCRIPT		0	0.00	$.............	....	402100
053-00	INSURANCE		4,635-	0.43	$.............	....	402200
054-00	LEGAL & AUDITING		827-	0.08	$.............	....	402300
055-00	WORKMENS COMP INSUR		200-	0.02	$.............	....	402400
056-00	BANK CHARGES		0	0.00	$.............	....	402500
057-00	GROUP INSURANCE		0	0.00	$.............	....	402600
059-00	REPAIRS & MAINT-BLDG		824-	0.08	$.............	....	402700
060-00	DEPRECIATION		0	0.00	$.............	....	402800
061-00	BAD DEBTS		0	0.00	$.............	....	402900
062-00	BAD DEBTS COLLECTION C		0	0.00	$.............	....	403000
064-00	INTEREST		560-	0.05	$.............	....	403100
067-00	DONATIONS		0	0.00	$.............	....	403200
068-00	AFTERSALE WARR-NEW EQ		4,000-	0.37	$.............	....	403300
069-00	AFTERSALE WARR-USED		2,000-	0.19	$.............	....	403400
070-00	AFTERSALE ADJ-SERVICE		4,000-	0.37	$.............	....	403500
071-00	MISCELLANEOUS EXP		2,010-	0.19	$.............	....	403600
*TOTAL EXPENSES		*	60,577-	5.61	....	....	
351-00	INTEREST EARNED		1,392	0.13	$.............	....	500103
352-00	CASH DISCTS EARNED		150	0.01	$.............	....	500203
353-00	FINANCE CHGS EARNED		0	0.00	$.............	....	500303

```
DM20ST                              BUDGET PROJECTION WORKSHEET
DEALER 7-99-9999-9               FISCAL YEAR ENDING  DECEMBER, 1979
```

ACCOUNT NO. SALES COST	A C C O U N T	...LAST YEAR... SALES	MARGIN	%	...YOUR PROJECTION FOR CURRENT YEAR... SALES COST MARGIN/EXPENSE %	
380-00	SUNDRY GAINS/LOSSES		0	0.00	$.............	500403
382-00	BAD DEBTS RECOVERED		0	0.00	$.............	500503
384-00	GAIN ON FIXED ASSET		0	0.00	$.............	500603
385-00	CASH OVER & SHORT		0	0.00	$.............	500703
386-00	OTHER COMMS & EARN		0	0.00	$.............	500803
387-00	INCOME CASE PROGRAM		0	0.00	$.............	500903
*TOTAL OTHER INC/DEDUCTIONS		*	1,542	0.14	....	
*NET PROFIT OR LOSS BEFORE TAXES		*	166,121	15.39	....	

| DESCRIPTION | . . . . S A L E S . . . | | | | * | . . . . M A R G I N S . . . . . | | | | |
	Y-T-D LAST YEAR	Y-T-D THIS YEAR	BUDGET	VARIANCE UNFAV-	*	Y-T-D LAST YEAR	% OF SALES	Y-T-D THIS YEAR	% CF SALES	BUDGET	VARIANCE UNFAV-
			NEW EQUIPMENT DEPARTMENT								
NEW TRACTORS		4C5,68C	275,000	130,680	*	55,702		73,215	18.C	275,000	201,784-
NEW IMPLEMENTS	21,65C	145,760	281,250	135,489-	*	951	4.4	28,272	19.4	281,250	252,977-
TOTAL NEW EQUIPMENT	21,65C	551,441	556,250	4,808-	*	56,654	261.7	101,487	18.4	556,250	454,762-
			USED EQUIPMENT DEPARTMENT								
USED TRACTORS	16,74C	250,967	122,915	128,052	*	645	3.9	55,134	22.C	122,915	67,780-
USED IMPLEMENTS	18,825	5,000	125,000	120,000-	*	1,667	8.9	2,500	50.0	125,000	122,500-
TOTAL USED EQUIPMENT	35,565	255,967	247,915	8,052	*	2,313	6.5	57,634	22.5	247,915	190,280-
TOTAL NEW AND USED	57,215	8C7,409	804,165	3,244	*	58,968	103.1	159,122	19.7	804,165	645,042-
			PARTS DEPARTMENT								
PARTS-CUSTOMER	32,625	147,847	66,665	81,182	*	9,788	30.0	31,692	21.4	66,665	34,972-
PARTS-INTERVAL	16,277	55,928	18,750	37,178	*	4,883	30.0	9,102	16.3	18,750	9,647-
TOTAL PARTS	48,907	203,776	85,415	118,361	*	14,672	30.0	40,794	2C.C	85,415	44,620-
			SERVICE DEPARTMENT								
SERVICE LAB-CUST	1C,505	60,200	46,000	14,200	*	1,277	12.2	28,858	47.9	46,000	17,141-
SERVICE LAB-INTERVAL		51,752	30,665	21,087	*		100.0	51,752	100.0	30,665	21,087
TOTAL SERVICE	1C,505	111,952	76,665	35,287	*	1,277	12.2	8C,61C	72.C	76,665	3,945
			RENTAL DEPT.								
RENTAL-NON-TAXABLE		151,950	83,330	68,620	*		100.0	24,45C	16.1	83,330	58,880-
TOTAL RENTAL DEPT		151,950	83,330	68,620	*			24,450	16.1	83,330	58,380-
			ALL DEPARTMENTS								
GRAND TOTAL	116,627	1275,088	1049,575	225,513	*	74,918	64.2	304,977	23.9	1049,575	744,597-
LESS TOTAL EXPENSE					*	45,508-	39.0	72,882-	5.7	105,535-	32,652
OPERATING PROFIT OR LOSS					*	29,410	25.2	232,C95	18.2	944,040	711,944-
			OTHER INCOME & DEDUCTIONS								
INTEREST EARNED					*			1,699	.1	2,C80	380-
CASH DISCTS EARNED					*	880	.8	150		6,250	6,099-
FINANCE CHGS EARNED					*	369	.3			625	625-
SUNDRY GAINS/LOSSES					*						
BAD DEBTS RECOVERED					*						
GAIN ON FIXED ASSET					*	8,625	7.4				
CASH OVER & SHORT					*	1-					
OTHER COMMS & EARN					*	53				80	80-
INCOME CASE PROGRAM					*	815	.7			830	830-
TOTAL OTHER INCOME-DEDUC					*	10,744	9.2	1,849	.1	9,865	8,015-
TOTAL NET PROFIT OR LOSS					*	40,154	34.4	233,945	18.3	952,905	719,959-

| DESCRIPTION | . . . . . S A L E S . . . . . | | | | * | . . . . . M A R G I N S . . . . . | | | | | |
	MAY LAST YEAR	MAY THIS YEAR	BUDGET	VARIANCE UNFAV-	*	MAY LAST YEAR	% OF SALES	MAY THIS YEAR	% OF SALES	BUDGET	VARIANCE UNFAV-
NEW EQUIPMENT DEPARTMENT											
NEW TRACTORS		60,000	55,000	5,000	*	4,200		12,000	20.0	55,000	43,000-
NEW IMPLEMENTS	2,000	20,000	56,250	36,250-	*	182	9.1	4,000	20.0	56,250	52,250-
TOTAL NEW EQUIPMENT	2,000	80,000	111,250	31,250-	*	4,382	219.1	16,000	20.0	111,250	95,250-
USED EQUIPMENT DEPARTMENT											
USED TRACTORS	3,100	40,000	24,583	15,417	*	247	8.0	10,000	25.0	24,583	14,583-
USED IMPLEMENTS	12,109	1,000	25,000	24,000-	*	1,164	9.6	500	50.0	25,000	24,500-
TOTAL USED EQUIPMENT	15,209	41,000	49,583	8,583-	*	1,411	9.3	10,500	25.6	49,583	39,083-
TOTAL NEW AND USED	17,209	121,000	160,833	39,833-	*	5,794	33.7	26,500	21.9	160,833	134,333-
PARTS DEPARTMENT											
PARTS-CUSTOMER	5,175	20,000	13,333	6,667	*	2,753	30.0	5,000	25.0	13,333	8,333-
PARTS-INTERNAL	1,900	5,000	3,750	1,250	*	570	30.0	5,000	100.0	3,750	3,750-
TOTAL PARTS	11,079	25,000	17,083	7,917	*	3,323	30.0	10,000	40.0	17,083	12,083-
SERVICE DEPARTMENT											
SERVICE LAB-CUST	1,671	10,000	9,200	800	*	29-	1.8	6,647	66.5	9,200	2,552-
SERVICE LAB-INTERNAL		10,000	6,133	3,867	*		100.0	10,000	100.0	6,133	3,867
TOTAL SERVICE	1,671	20,000	15,333	4,667	*	29-	1.8	16,647	83.2	15,333	1,314
RENTAL DEPT.											
RENTAL-NON-TAXABLE		30,000	16,666	13,334	*		100.0	4,500	15.0	16,666	12,166-
TOTAL RENTAL DEPT		30,000	16,666	13,334	*			4,500	15.0	16,666	12,166-
ALL DEPARTMENTS											
GRAND TOTAL	29,960	196,000	209,915	13,915-	*	9,088	30.3	57,647	29.4	209,915	157,267-
LESS TOTAL EXPENSE					*	5,519-	18.4	12,288-	6.3	21,107-	8,818
OPERATING PROFIT OR LOSS					*	3,569	11.9	45,358	23.1	188,308	148,449-
OTHER INCOME & DEDUCTIONS											
INTEREST EARNED					*			304	.2	416	111-
CASH DISCTS EARNED					*	90	.3			1,250	1,250-
FINANCE CHGS EARNED					*	83	.3			125	125-
SUNDRY GAINS/LOSSES					*						
BAD DEBTS RECOVERED					*						
GAIN ON FIXED ASSET					*						
CASH OVER & SHORT					*						
OTHER COMMS & EARN					*	58-	.2			16	16-
INCOME CASE PROGRAM					*	69-	.2			166	166-
TOTAL OTHER INCOME-DEDUC					*	46	.2	304	.2	1,973	1,668-
TOTAL NET PROFIT OR LOSS					*	3,615	12.1	45,663	23.3	190,781	150,117-

| DESCRIPTION | . . . . MONTH OF MAY . . . . . | | | * | . . . . % YEAR-TO-DATE . . . . | | | | | | VARIANCE |
	LAST YEAR	THIS YEAR	BUDGET	*	LAST YEAR	% SALES	THIS YEAR	% SALES	BUDGET	% SALES	UNFAV-
SALARIES-MGRS	1,200	1,383	2,816	*	6,300		8,700		14,080		5,379
SALARIES-OFFICE EMPL		752	500	*	509		3,763		2,500		1,263-
SALARIES-SALESMEN	100	329	1,250	*	750		2,730		6,250		3,519
SALARIES-PARTSMAN	619	3,102	2,816	*	3,260		16,594		14,080		2,514-
ELECTRONIC ACCTG SER			200	*	917		517		1,000		482
RENT & LEASE EXP	450	1,000	500	*	2,250		6,478		2,500		3,978-
LEASE EQUIP				*							
UTILITIES		500	583	*	730		2,806		2,915		108
TELEPHONE	202	200	250	*	820		1,000		1,250		250
OFFICE & STORE SUPPLY	16	50	41	*	655		593		205		388-
EMPLOYEE BENEFITS		160	250	*			800		1,250		450
FREIGHT MISC		50		*			651				651-
SHOP SUPPLY & EQ REP	77	25	666	*	774		623		3,330		2,706
CAR & TRUCK EXP	5	25	1,125	*	1,212		144		5,625		5,481
GAS & OIL		10		*	1,774		264				264-
TRAVEL & SALES		50	166	*	192		250		830		580
ADVERTISING	440	500	666	*	1,986		3,161		3,330		169
PAYROLL-TAXES	142		1,416	*	821		594		7,080		6,485
TAXES-BUS PPTY			250	*	481				1,250		1,250
OTHER TAX & LIC			1,250	*	1,048				6,250		6,250
DUES & SUBSCRIPT			16	*	25				80		80
INSURANCE	60	1,000	1,250	*	3,854		5,635		6,250		614
LEGAL & AUDITING			500	*	1,086		827		2,500		1,672
WORKMENS COMP INSUR		50		*	1,048		250				250-
BANK CHARGES	25		1,000	*	135				5,000		4,999
GROUP INSURANCE	216		691	*	1,084				3,455		3,455
REPAIRS & MAINT-BLDG	40	100	83	*	581		924		415		509-
DEPRECIATION			625	*					3,125		3,125
BAD DEBTS			41	*	11				205		205
BAD DEBTS COLLECTION C				*							
INTEREST	660		1,000	*	3,394		560		5,000		4,439
DONATIONS	10		41	*	28				205		205
AFTERSALE WARR-NEW EQ	816	1,000	708	*	3,796		5,000		3,540		1,460-
AFTERSALE WARR-USED		500	291	*	3,784		2,500		1,455		1,045-
AFTERSALE ADJ-SERVICE		1,000	50	*	21		5,000		250		4,750-
MISCELLANEOUS EXP	436	500	66	*	2,191		2,510		330		2,180-
TOTAL EXPENSE	5,519	12,288	21,107	*	45,508		72,882		105,535		32,652

DESCRIPTION	. . . . . S A L E S . . .				*	. . . . . M A R G I N S					
	Y-T-D 2-YRS AGO	Y-T-D LAST YEAR	Y-T-D CURR YEAR	VARIANCE UNFAV-	*	Y-T-D 2-YRS AGO	Y-T-D LAST YEAR	% OF SALES	Y-T-D CUFR YEAR	% OF SALES	VARIANCE UNFAV-
NEW EQUIPMENT DEPARTMENT											
NEW TRACTORS	52,5CC		405,680	405,680	*	1,385-	55,702		73,215	18.0	17,512
NEW IMPLEMENTS	3,600	21,650	145,760	124,110	*	9,843-	951	4.4	28,272	19.4	27,320
TOTAL NEW EQUIPMENT	56,100	21,650	551,441	529,791	*	11,228-	56,654	261.7	101,487	18.4	44,832
USED EQUIPMENT DEPARTMENT											
USED TRACTORS	18,682	16,740	250,967	234,227	*	11,348	645	3.9	55,134	22.0	54,488
USED IMPLEMENTS	28,544	18,825	5,000	13,825-	*	4,424	1,667	8.9	2,500	50.0	832
TOTAL USED EQUIPMENT	47,226	35,565	255,967	220,402	*	15,773	2,313	6.5	57,634	22.5	55,321
TOTAL NEW AND USED	103,326	57,215	807,409	750,193	*	4,544	58,968	103.1	159,122	19.7	100,154
PARTS DEPARTMENT											
PARTS-CUSTOMER	28,372	32,629	147,847	115,218	*	8,511	9,788	3C.C	31,692	21.4	21,903
PARTS-INTERNAL	4,306	16,277	55,928	39,650	*	1,292	4,883	30.0	9,102	16.3	4,218
TOTAL PARTS	32,679	48,907	203,776	154,869	*	9,803	14,672	30.C	4C,794	20.0	26,122
SERVICE DEPARTMENT											
SERVICE LAB-CUST	8,106	10,505	60,200	49,694	*	1,111-	1,277	12.2	28,858	47.9	27,581
SERVICE LAB-INTERNAL			51,752	51,752	*			100.0	51,752	100.0	51,752
TOTAL SERVICE	8,106	10,505	111,952	101,446	*	1,111-	1,277	12.2	8C,610	72.0	79,333
RENTAL DEPT.											
RENTAL-NON-TAXABLE			151,950	151,950	*			1CC.C	24,450	16.1	24,450
TOTAL RENTAL DEPT			151,950	151,950	*				24,450	16.1	24,450
ALL DEPARTMENTS											
GRAND TOTAL	144,112	116,627	1275,088	1158,460	*	13,237	74,918	64.2	304,977	23.9	230,059
LESS TOTAL EXPENSE					*	37,615-	45,508-	39.C	72,882-	5.7	27,373-
OPERATING PROFIT OR LOSS					*	24,377-	29,41C	25.2	232,095	18.2	202,685
OTHER INCOME & DEDUCTIONS											
INTEREST EARNED					*				1,699	.1	1,699
CASH DISCTS EARNED					*	987	880	.8	150		729-
FINANCE CHGS EARNED					*	479	369	.3			369-
SUNDRY GAINS/LOSSES					*						
BAD DEBTS RECOVERED					*						
GAIN ON FIXED ASSET					*	5,362	8,625	7.4			8,625-
CASH OVER & SHORT					*		1-				1
OTHER COMMS & EARN					*		53				53-
INCOME CASE PROGRAM					*		815	.7			815-
TOTAL OTHER INCOME-DEDUC					*	6,830	10,744	5.2	1,849	.1	8,894-
TOTAL NET PROFIT OR LOSS					*	17,547-	40,154	34.4	233,945	18.3	193,791

DESCRIPTION	. . . . S A L E S				*	. . . . M A R G I N S					
	MAY 2-YRS AGO	MAY LAST YEAR	MAY CURR YEAR	VARIANCE UNFAV-	*	MAY 2-YRS AGO	MAY LAST YEAR	% OF SALES	MAY CURR YEAR	% OF SALES	VARIANCE UNFAV-
NEW EQUIPMENT DEPARTMENT											
NEW TRACTORS			60,000	60,000	*	4,098-	4,200		12,000	20.0	7,800
NEW IMPLEMENTS		2,000	20,000	18,000	*	36-	182	9.1	4,000	20.0	3,817
TOTAL NEW EQUIPMENT		2,000	80,000	78,000	*	4,135-	4,382	215.1	16,000	20.0	11,617
USED EQUIPMENT DEPARTMENT											
USED TRACTORS	5,200	3,100	40,000	36,900	*	103	247	8.C	1C,000	25.0	9,752
USED IMPLEMENTS	2,644	12,109	1,000	11,109-	*	558	1,164	9.6	500	50.0	664-
TOTAL USED EQUIPMENT	7,844	15,209	41,000	25,790	*	661	1,411	9.3	1C,500	25.6	9,088
TOTAL NEW AND USED	7,844	17,209	121,000	103,790	*	3,473-	5,794	33.7	26,500	21.9	20,705
PARTS DEPARTMENT											
PARTS-CUSTOMER	5,567	9,179	20,000	10,820	*	1,670	2,753	30.C	5,000	25.0	2,246
PARTS-INTERNAL	130	1,900	5,000	3,099	*	39	570	3C.0			570-
TOTAL PARTS	5,698	11,079	25,000	13,920	*	1,709	3,323	3C.C	5,000	20.0	1,676
SERVICE DEPARTMENT											
SERVICE LAB-CUST	1,398	1,671	10,000	8,328	*	308-	29-	1.8	6,647	66.5	6,677
SERVICE LAB-INTERNAL			10,000	10,000	*						
TOTAL SERVICE	1,398	1,671	20,000	18,328	*	308-	29-	1.8	6,647	33.2	6,677
RENTAL DEPT.											
RENTAL-NON-TAXABLE			30,000	30,000	*				4,500	15.0	4,50C
TOTAL RENTAL DEPT			30,000	30,000	*				4,500	15.0	4,500
ALL DEPARTMENTS											
GRAND TOTAL	14,94C	29,960	196,000	166,039	*	2,072-	9,088	3C.3	42,647	21.8	33,558
LESS TOTAL EXPENSE					*	8,558-	5,519-	18.4	12,288-	6.3	6,768-
OPERATING PROFIT OR LOSS					*	10,631-	3,569	11.9	30,358	15.5	26,789
OTHER INCOME & DEDUCTIONS											
INTEREST EARNED					*				304	2.0	304
CASH DISCTS EARNED					*	323	90	.3			90-
FINANCE CHGS EARNED					*	124	83	.3			83-
SUNDRY GAINS/LOSSES					*						
BAD DEBTS RECOVERED					*						
GAIN ON FIXED ASSET					*						
CASH OVER & SHORT					*						
OTHER COMMS & EARN					*		58-	.2			58
INCOME CASE PROGRAM					*		69-	.2			69
TOTAL OTHER INCOME-DEDUC					*	447	46	.2	304	2.0	258
TOTAL NET PROFIT OR LOSS					*	10,183-	3,615	12.1	30,663	205.2	27,047

DESCRIPTION	. . . . MONTH OF MAY			*	. . . . YEAR-TO-DATE						VARIANCE
	2-YRS AGO	LAST YEAR	CURR YEAR	*	2-YRS AGO	% SALES	LAST YEAR	% SALES	CURR YEAR	% SALES	UNFAV-
SALARIES-MGRS		1,200	1,383	*			6,300	5.4	8,700	.7	2,400-
SALARIES-OFFICE EMPL	186		752	*	1,033	.7	505	.4	3,763	.3	3,254-
SALARIES-SALESMEN	150	100	329	*	800	.6	750	.6	2,730	.2	1,980-
SALARIES-PARTSMAN	711	619	3,102	*	3,700	2.6	3,260	2.8	16,594	1.3	13,334-
ELECTRONIC ACCTG SER				*	389	.3	917	.8	517		400
RENT & LEASE EXP	450	450	1,000	*	2,250	1.6	2,250	1.9	6,478	.5	4,228-
LEASE EQUIP				*							
UTILITIES	953		500	*	2,662	1.8	730	.6	2,806	.2	2,075-
TELEPHONE	139	202	200	*	737	.5	820	.7	1,000	.1	179-
OFFICE & STORE SUPPLY		16	50	*	204	.1	655	.6	593		62
EMPLOYEE BENEFITS			160	*	109	.1			800	.1	800-
FREIGHT MISC			50	*					651	.1	651-
SHOP SUPPLY & EQ REP	140	77	25	*	527	.4	774	.7	623		151
CAR & TRUCK EXP	63	5	25	*	1,279	.9	1,212	1.0	144		1,068
GAS & OIL	1,202		10	*	2,439	1.7	1,774	1.5	264		1,510
TRAVEL & SALES			50	*	700	.5	192	.2	250		57-
ADVERTISING	136	440	500	*	1,336	.9	1,986	1.7	3,161	.2	1,174-
PAYROLL-TAXES	206	142		*	1,033	.7	821	.7	594		226
TAXES-BUS PPTY				*	506	.4	481	.4			481
OTHER TAX & LIC	198			*	751	.5	1,048	.9			1,048
DUES & SUBSCRIPT				*	39		25				25
INSURANCE	2,524	60	1,000	*	4,691	3.3	3,854	3.3	5,635	.4	1,780-
LEGAL & AUDITING				*	200	.1	1,086	.9	827	.1	258
WORKMENS COMP INSUR	198		50	*	751	.5	1,048	.9	250		798
BANK CHARGES	23	25		*	199	.1	135	.1			134
GROUP INSURANCE		216		*			1,084	.9			1,084
REPAIRS & MAINT-BLDG		40	100	*	180	.1	581	.5	924	.1	342-
DEPRECIATION				*							
BAD DEBTS				*			11				11-
BAD DEBTS COLLECTION C				*							
INTEREST	617	660		*	7,472	5.2	3,394	2.9	560		2,834
DONATIONS		10		*	35		28				28
AFTERSALE WARR-NEW EQ	34	816	1,000	*	125	.1	3,796	3.3	5,000	.4	1,203-
AFTERSALE WARR-USED	129		500	*	865	.6	3,784	3.2	2,500	.2	1,284
AFTERSALE ADJ-SERVICE	41		1,000	*	581	.4	21		5,000	.4	4,978-
MISCELLANEOUS EXP	449	436	500	*	2,257	1.6	2,191	1.9	2,510	.2	318-
TOTAL EXPENSE	8,558	5,519	12,288	*	37,615	26.1	45,508	38.8	72,882	5.5	27,373-

SALES PRICE BOOK - USED TRACTORS & EQUIPMENT   DATE 9/01/81   PAGE   3

STOCK#	MAKE	MODEL	SPECS / EXTENDED SPECS	SR #	ID	LOC	DATE	DISC	SPECIAL	PRICE	COST
********** A-TRAC-FORD *****											
UT5334	FORD	5000	69,4080,8-S,PS-Y,61-N,RE	OT6949	G	MV	01-81		3	$8950.00	5856.98
********** A-TRACTOR *****											
UT5439	FORD	5000	71,4952,8-S,PS-Y	C787666		LN	06-81	7.0	4	$9549.00	6138.00
UT5442	FORD	6600	EA214,A3,DF,E8,GF,K3,L7,N3	C876555		MV	06-81	7.0	4	$10500.50	7298.00
********** M-BALER *****											
UE0484	JOHN DEERE	14T	TWINE P.T.O.	54212	G	LN	01-81	.3		$1395.00	1109.25
********** T-CHOPPER *****											
UE0174	INT'L	350	HARVESTER W/WIND P/U	651	G	LN	01-81	.3		$1595.00	1253.00
********** TA-HEADS *****											
UE5189	NEW HOLLAND	717R	1 ROW CORNHEAD	122369	G	MV	01-81		3	$595.00	900.00
UE0715	NEW HOLLAND	770W	WINDROW PICKUP	271405		LY	01-81			$.00	135.80
********** TLB *****											
UT5576	FORD	TLB	GC315M,DF,GH,J5,LA,Q8,Y8	C765444		LN	06-81		4	$17977.00	12202.00
********** U-FLAIL *****											
UE5857	NEW HOLLAND	36				MV	01-81			$795.00	420.00
********** V-BLOWER *****											
UE0268	NEW HOLLAND	27	BLOWER	213271	R	LN	01-81	.3		$995.00	786.68

```
WHOEVER TRACTOR COMPANY
ANYWHERE, USA          A     WHOLE GOODS INQUIRY                                    DATE   8/13/81
                                                                                           PAGE  1

STOCK#  MAKE         MODEL  DESCRIPTION        SPECS                    SR #     ID LOC DATE DISC WAR   SPECIAL    PRICE
NH5619 NEW HOLLAND  315T   A-BALER      3JTPPHD,SS,FLT-TIRES,3P-CHUTE  454863    S MV 01-31      4               $7501.00
SOLD...
                         SOLD TO    LEROY GRADY                                                        SOLD FOR    7501.00
                         SALE DATE 06/05/81                                                               COST     6231.05
                         DOCUMENT  C45601
                         TRADED ON                                                                     MARGIN ..........  16.93 %

TRANS...
         P    1/24/81  32687        O  8124                                                                        6231.05
         *    6/05/81  C45601       C  8124        A-BALER       G50525                                            6231.05-
         *    6/05/81  C45601       U  A424        A-BALER       G50525                                            6231.05
                                                                                            SUBTOTAL              6231.05

WASHOUT...                                      INV.   DATE        SOLD
                                        STOCK#  DAYS   SOLD         FOR        COST    MARGIN    TRADE ON   WASHOUT TREE
         NEW HOLLAND  315T  A-BALER     NH5619  132   6/05/81     7501.00    6231.05   15.93 %              NH5619
         JOHN DEERE   14T   SALE W.TRADE UE0484 202   8/13/81     1395.00    1109.25   20.48 %   NH5619     UE0484
         NEW HOLLAND  27                 UE0585   5   8/13/81      475.00     295.00   37.89 %   UE0484      UE0585
                                WASHOUT TOTALS                    9371.00    7635.30   18.52 %
```

| STOCK | N/U | SERIAL | MODEL | DATE | MEMO | SEQ | DESCRIPTION | RETAIL | SOLD TO | SOLD |
NO	/R UNITS	NO	NO			NO				DATE
						080	3626 6405 DOZER			
						090	9181 DRAWBAR			
I05117	R	1 C329550T	JD450C	06/22/79 72	T76638	010	3303T JD450C DOZER	34239		
						020	0001 4219CT02 DSL ENG			
						030	1205 23GPM PUMP			
						040	4525 16IN GROUSER HARD			
						050	7013 95AMP BATTERY			
						060	8401 BOTT GRD USE W/DBR			
						070	8570 NO CAB OR ROPS			
						030	8626 6405 DOZER			
						090	9181 DRAWBAR			
I05118	N	1 A367479T	JD450C	06/26/81 53	W18334	010	3313T JD450C LOADER	43645		
						020	0001 4219TCD01 DSL ENG			
						030	7014 95AMP BATTERY			
						040	8570 NO CAB OR ROPS			
						050	9601 1 1/4 YD BUCKET			
						060	9616 RETURN TO DIG			
I10023	N	1- C369121T	JD550	09/02/81 14	V03428	010	3863T JD550 DOZER	51444		
						020	0001 4276TT DSL ENG			
						030	1205 23GPM HYD PUMP			
						040	2150 DUAL LEVER STEER			
						050	4525 16IN GROUSER			
						060	7013 95AMP BATTER W/C			
						070	8401 BTM GRD W/DBR			
						030	8570 NO CAB OR ROPS			
						090	8626 6405 DOZER			
						100	9181 DRAWBAR			
I10024	N	1 C369122T	JD550	03/28/81 14	V64428	010	3863T JD550 DOZER	51444		
						020	0001 4276TT DSL ENG			
						030	1205 23GPM HYD PUMP			
						040	2150 DUAL LEVER STEER			
						050	4525 16IN GROUSER			
						060	7013 95AMP BATTERY W/C			
						070	8401 BTM GUARD W/DBR			
						080	8570 NO CAB OR ROPS			
						090	8626 6405 DOZER			
						100	9181 DRAWBAR			
I10025	N	1 C366351T	JD550	05/29/81 04	V79492	010	3863T JD550 DOZER	49503		
						020	0001 4276TT DSL ENG			
						030	1205 23GPM HYD PUMP			
						040	2150 DUAL LEVER STEER			
						050	4525 16IN GROUSER			
						060	7013 95AMP BATTERY W/C			

A565DR01
DEALER 3-04-0099-3-01

A G E D   W H O L E   G O O D S   I N V E N T O R Y
COST AS OF 11/11/81

STOCK DESCRIPTION NO	N U	0-1 MONTH	2-3 MONTHS	4-6 MONTHS	7-9 MONTHS	10-12 MONTHS	13-18 MONTHS	OVER 18	FLR PLAN DUE DATE
I05112 3303T 450C DOZER	N		35127.73						10/15/82
I05113 3803T 450C DOZER	N			33850.53					03/15/81
I05114 3303T 450C DOZER	N			33850.53					03/15/82
I05115 3803T 450C DOZER	N			33350.53					03/15/82
I05116 3303T JD450C DOZER	N				33850.53				12/15/81
I05117 3603T JD450C DOZER	R							27336.67	03/15/80
I05118 3813T JD450C LOADER	N			35483.81					06/15/82
I10023 3863T JD550 DOZER	N		41824.30						06/15/82
I10024 3863T JD550 DOZER	N		41824.46						10/15/82
I10025 3363T JD550 DOZER	N			40294.97					03/15/82
I10026 3363T JD550 DOZER	R						38046.09		08/15/81
I20015 3373T	N				14296.21				12/15/81
I25043 4013T 310A BKHOE LDR	N		28420.94						06/15/82
I25044 4013T 310A BKH LDR	N		28420.94						10/15/82
I25045 4013 JD310A BKH LDR	N					26336.56			11/15/81
I27050 BU10103 12 INCH BCKTT	N					340.33			10/15/81
I27051 BU10103 12 INCH BUCKET	N					340.34			10/15/81
I27052 BU10618 16" BUCKET	N							395.29	02/15/80
I27053 BU10618 16" BUCKET	N							395.29	02/15/80
I27054 BU10396 16 INCH BUCKET	N					376.64			10/15/81
I27055 AT37722 18 INCH BUCKET	N					542.18			10/15/81
I27056 AT37729 18 INCH BUCKET	N					420.50			10/15/81
I27057 AT37229 18 INCH BUCKET	N					420.50			10/15/81
I27058 BU10620 24 INCH BUCKET	N				839.37				11/15/81
I27059 AT37725 36 INCH BUCKET	N					642.38			10/15/81
I29050 AT43152 CANOPY	N					909.17			10/15/81
I29051 AT43152 CANOPY	N					909.17			10/15/81
I29052 AT43151 CANOPY	N			1106.71					02/15/82
I29053 AT43151 CANOPY	N		1143.31						06/15/82
I29054 AT43151 CANOPY	N		1143.31						06/15/82
I29056 AT42733 CAB BL	N		2861.60						05/15/82
I30067 3583T JD410 LDR/BKH	N			30690.35					03/15/82
I30068 3583T 410 BKHOE LDR	N			30690.35					03/15/81
I30069 3583T JD410 BCK LDR	N						23735.34		03/15/81
I30070 3583T JD410 BCK LDR	R						23735.34		07/15/81
I50010 3833T JD480B FORKLIFT	N						21527.33		10/15/81
I50011 3833T JD480B FORKLIFT	N					22102.70			09/15/81
I60020 AT51555 9300 HOE	R						8850.17		06/15/81
I60021 AT51555 9300 HOE	N			9490.35					03/15/82
I65045 AT43936 9550 HOE	N				3583.26				12/15/81
I65046 AT43936 9550 HOE	N				3309.52				12/15/81
I65047 AT43936 9550 HOE	N				3306.24				02/15/82
			180756.64	249308.13	74185.13	53340.47	125894.27	28627.25	

22210041 DEALER — STORE 1 01 — MFR. GA — REPORT LOCATOR

NFPEDA
NATIONAL FARM AND POWER EQUIPMENT DEALERS ASSOCIATION

CYCLE NO. 477 — PAGE 25

CUSTOMER: SAMPLE IMPLEMENT CO. / 100 MAIN STREET / ST. LOUIS, MO. 63127
CUST. NO. 1942 — CYCLE NO. 477 — PAGE 25

22210041 DEALER — 1 STORE — GA MFR.

INDICATORS / PRICING & LOCATION / INVENTORY STATUS

Indicators

STAT	3/4	DATE COUNT IN DATE	P M S	PACK MULT SETS	FREEZE MIN	FREEZE MAX	SEASON BEG/END	YEAR TO-DATE PREV DATE	LINE NO.
		16			6	8		38	017
		56			4	14		18	021
		56			12	41		13	035
A		16			0	1		1	043
D		ND5			0	1		1	052
									069
M		N36			7	20		43	076
									088
									090
N								3	101
S		36	P	3P		869			114
									122
A		X76			0	1			132
R		D5			10	19		62	146
									159
		26S	S	8	24	32		192	161
		N16M	M	50	68	144		360	177
		N16P	P	75	89	367		476	180
									197
A		85			0	1			201
T									214
		76			0	1		8	223
									236
		66S	S	8	32	72		136	247
									255
		66	P	OP		1		10	266
									279
									281
									290
		N85			15	24		86	300
		D5			6	8			315
									329
									332
		26			6	9		34	342
									351
		26			4	8		17	364
M		16P100			59	254		353	374
N								1	385
		00	T	10T		10			397

Sales History — Current Year Sales

LINE NO.	MO 1 JUL	MO 2 JUN	MO 3 MAY	MO 4 APR	MO 5 MAR	MO 6 FEB	MO 7 JAN	MO 8 DEC	MO 9 NOV	MO 10 OCT	MO 11 SEP	MO 12 AUG	MO NO SALE	P FREEZE COST DIV
017	3	4	1	2		5	3	2	3	5	5	3		
021	1	5		2	3		1	1	2	3	3	4		
035	3	6	10	8	5	7	4	3	3	4	7	12		
043													14	
052													11	
076	1	3	4	2	3	1	4	2	5	4	7	6		
090					11								5	
101									3					
114	9	6				1								
132													34	
146	8	2	6	3	5	8	4	1	5	4	7	7		
161	16	24	16	8	16	8	8	8	16	16	24	32		
177	22	41	39	46	28	40	32	11	18	26	59	46		
180	26	43	49	53	46	48	56	20	34	38	47	74		P
201													53	
214													10	
223			2	1	2		1						2	
247	16	40	8	32	16	24	16		8	8	8	16		
266	1		2		1		1					1		
300	14	12	8	10	7	4	2	3	4	6	8	12		
315	2	5	6	3	2	4	3	1	1	2	2	4		
342	6	3	1	5	4	2	1	1	2	4	5	2		
364	1	1	2	1	3	1	2			7301				
374	23	40	29	24	31	27	12	14	21	29	46	58		P
385													8	
397			2	4										

Pricing & Location / Inventory Status

LINE NO.	2ND PREV YEAR	LAST ACTIVE DATE	EXTENDED VALUE	ALSO STOCKED EXTRA INFO	BIN LOCATION	PART NUMBER & DESCRIPTION	ON ORDER	ORDER NO.	ON HAND
017	14536	19378	21316		B160	4D1636 SHAFT			7
021	64	85	9423		A460	4F0886 BUSHING			11
035	37	48	533		AA36	2K0190 FLANGE			35
043	1601	2134	25462		AE159	6L4677 REGULATOR			1
052	315	480	598		AE120	4M4419 LOCK			1
069				RECORD KEY NUMBER IN FILE FOR CUSTOMER			REF1	***/DEL	
076	65	87	95		C120	4M8808 GASKET			16
088	13212	16515	1816723		******	5M7216 BLOCK			
090	15	20	22		N145	7M3802 PLUG	**DELETED***		
101	186	250	275		N-STK	7M4102 FLANGE			
114	164	185	20423		V134	7M4613 CAP	6		2
122							6	467	
132	6084	7605	8366		C147	7M8011 RADIATOR			7
146	240	300	330		NE140	8M3614 BAFFLE			6
159				NEW NO. 6B2437	OLD	8M3614	PART CHANGED		
161	1246	1652	18172		B143	8M4103 PISTON			22
177	07	09	1023		D436	8M5642 SHIM			73
180	03	05	0623		436B	8M6016 LOCKWASHER	375		60
197							375	467	
201	1560	1950	2145		E246	8M7214 MIRROR			3
214					R-BIN	9M3604 FAN ASSEMBLY			1
223	1842	2216	2438		G160	9M4102 SEAT	1		0
236							1	277	
247	36	55	6023		D438	9M4316 SPARK PLUG			48
255				NEW NO. 4A1742	OLD	9M7214	PART CHANGED		
266	7600	9576	10534		FLOOR	9M7516 COUNTERWEIGHT			1
279				USE ON MF MODEL 436. EACH WEIGHT APPROX			REF1	118/DEL	
281				500 LB. ORDER AS SOLD. ADD FREIGHT CHARGE			REF2	118/DEL	
290				OF $12.75 TO INVOICE.			REF3	118/DEL	
300	510	760	83623		EE460	9M7602 END BIT			16
315	1601	2134	2347		AC76	9M7815 ROLLER	7		1
329								3B467	
332								4B267	
342	1254	1673	1845		43180	4R2361 PIN ASSEMBLY	108		7
351								108C277	
364	520	650	7152		C512	5R2780 FUEL PUMP			7
374	15	21	2423		2512	6R6207 WASHER			143
385	4120	5120	5665		N-STK	6T2436 FLY WHEEL			
397	415	519	5712		C472	2V9412 SHOCK			16

NFPEDA

NATIONAL FARM AND POWER EQUIPMENT DEALERS ASSOCIATION

REPORT SUGGESTED STOCK ORDER

DEALER	22210041	STORE	1 46	MFR.	GA
CYCLE NO.	477	PAGE	10		

CUSTOMER	SAMPLE IMPLEMENT CO. 100 MAIN STREET ST. LOUIS, MO. 63127
CUST. NO.	1942
CYCLE NO.	477
PAGE	10 · GA · MFR.
DEALER	22210041 · STORE 1

INDICATORS / SALES HISTORY

Part No.	STAT	CLASS	P/M/S	PACK	MIN	MAX	YR-TO-DATE / PREV DATE	MO 1	MO 2	MO 3	MO 4	MO 5	MO 6	MO 7	MO 8	MO 9	MO 10	MO 11	MO 12	MO NO SALE
463146	R	32			7	14	25	6	2	3	5	3	4	2						
463146							43	7	3	1	3	4	6	5	4	2	1	4	3	
463150	N	41			3	11	11	3	2	1	1	2		2						
463150							18	1	1	3	2	2	1		3	1		1	3	
463162	R	D2			3	5	8	2	2	1		1	2							
463162							21	2	2	1	3	1	1	2		1	1	3	4	
463649	N	30	P	100	93	274	300	38	42	50	45	50	55	20						
463649							528	46	32	56	40	38	36	48	56	70	38	42	26	
470011	N	42			7	10	22	3	2	1	3	3	5	5						
470011							36	4	4	2	2	1	1	3	4	4	5	2	4	
471786	R	01			3	5	6	2	1		1		2							
471786							20	4	2	2		2			2		3	2	3	
477663	SR	38			27 / 12	2798 / 18	5	3	1	1										3
477663							71			1			1		4	12	16	21	15	
564381	R	96			3	5	11	3	1	2	1	1	3							
564381							21	2	4	4		1	2	3	2	1		2		
614305	N	72			12	21	45	5	4	4	5	9	10	8						
614305							71	8	5	6	6	8	7	4	5	6	5	5	6	
694638	R	59			1	4	3				1			2						
694638							21	1	3	6	5	1			2	1			2	
763842	R	80			3	5	8		2	3		1		2						
763842							25	6	4	5	1				3	2	2	1	1	
764396	R	42			0	1	2						1	1						
764396							5	1		2				1					1	
774369	N	74	P	100	103	289	321	56	63	21	49	53	49	30						
774369							607	48	50	43	61	40	39	46	52	38	46	68	76	
775214	R	82			6	19	18	4	1	3	2	4	3	1						
775214							43	4	4	4	3	2	5	5	2	3	5	4	2	
783291	R	D9	S	8	24	32	80	8	8	16	8	16	16	8						
783291							96		8	16	16	8		8		16	8	8	8	
840886	R	51	M	2	3	12	11	2			2	4		3						
840886							23	4	2	1	1	2	3	1	2		3	1	3	
851267	N	N1			1	2	3	1					1	1						
851267							4		1				1			1			1	
864201	R	54		T 10 / 4	T 20 / 8	24					8	4	12							

PRICING & LOCATION / INVENTORY STATUS

	2ND PREV YEAR / LAST ACTIVE DATE	EXTENDED VALUE	BIN LOCATION	PART NUMBER & DESCRIPTION	ON ORDER	ORDER NO.	ON HAND	LINE NO.	P FREEZE COST DIV
211	36277	16882	AE130	463146 BRACKET	8	477	6		
78	16257	702	D4630	463150 GASKET	9	477	2		
1475	16467	44253	CE460	463162 BRAKE	3	477	2		
4755	10277	1425	AA136	463649 BOLTS	PK 3	477	69		P
1075	36277	4300	ZA460	470011 LINER	4	477	1	3B267 / 2B257	
845	10257	253523	12046	471786 PIN ASSEMBLY	3	477	2		
	*** SURPLUS 5 BRANCH 2 ***								
580	63147	6960	CA463	477663 GUIDE	12	477	15		
875	15277	1750	YARD	564381 TRACK LINK	2	477	1	2 467	
382	68277	3438	AG46	614305 SEAL	9	477	1	7 467 / 4B267	
198	3347	594	BA943	694638 LEVER	3	477	1		
1236	30277	3708	D9640	763842 ROD	3	477	2		
4680	5467	4680	AE113	764396 FUEL TANK	1	477		1C267	
5005	75277	1000	GG463	774369 O-RING	PK 2	477	90		P
65	41467	845	AB204	775214 GASKET	13	477	6		
1246	120277	19936	BC416	783291 PISTON	16	477	16		
64	25277	640	CA302	840886 BUSHING	10	477	2		
1601	2277	1601	CB125	851267 REGULATOR	1	477		1 267	
415	277	4980	AF327	864201 SHOCK	12	477	8		
	TOTAL	261143							

NFPEDA

NATIONAL FARM AND POWER EQUIPMENT DEALERS ASSOCIATION

DEALER 22210041	CYCLE NO. 477	CUSTOMER SAMPLE IMPLEMENT CO.
STORE 1 03	PAGE 1	100 MAIN STREET
MFR. GA	REPORT STOCK ANALYSIS	ST. LOUIS, MO. 63127
		CUST. NO. 1942 CYCLE NO. 477 PAGE 1 GA
		DEALER 22210041 STORE 1 MFR. GA

INDICATORS	SALES HISTORY	PRICING & LOCATION	INVENTORY STATUS

Column headers: MIN DIV / SORT / CLASS CODES 3 4 / STAT / DATE COUNT IN DATE / P M / PACK MULT SETS / FREEZE / MIN / FREEZE / MAX / SEASON / BEG END / YEAR TO-DATE PREV DATE / MO 1–MO 12 / MO NO SALE / P FREEZE COST DIV / 2ND PREV YEAR / LAST ACTIVE DATE / EXTENDED VALUE / ALSO STOCKED EXTRA INFO / BIN LOCATION / PART NUMBER & DESCRIPTION / ON ORDER / ORDER NO. / ON HAND / LINE NO.

	QUANTITIES			PERCENTAGES		
	PIECES	PARTS	DOLLARS	PIECES	PARTS	DOLLARS
INVENTORY QUANTITIES AND VALUES						
BEGINNING STOCK PARTS	47010	7835	94020.87			
ENDING STOCK PARTS	43652	7860	101032.48			
NON-STOCK PARTS		1175			15.0	
STOCK EXCHANGE						
STOCK SALES	3996	769	9663.10	9.2	9.8	9.6
NON-STOCK SALES	542	147	2710.00	1.2	1.9	2.8
FACTORY DIRECT SALES						
STOCK LOST SALES	21	10	46.41	.0	.1	.0
NON-STOCK LOST SALES	38	36	63.19	.1	.0	.1
CREDIT RETURNS	16	8	49.28	.0	.1	.0
TRANSFERS IN	10	4	67.23	.1	.2	.1
TRANSFERS OUT	15	8	97.75	.0	.1	.1
NEW PARTS THIS PROCESS	73	35	200.12	.2	.4	.2
DELETIONS THIS PROCESS		10			.1	
NEW ORDERS	360	115	987.15	.8	1.5	1.0
OUTSTANDING ORDERS	541	286	1510.27	1.2	3.6	1.5
BACK ORDERS	467	121	981.32	1.1	1.5	1.0
STOCK RECEIPTS	641	216	1236.19	1.5	2.7	1.2
EMERGENCY RECEIPTS	296	79	846.98	.7	1.0	.8
VENDOR RECEIPTS	150	15	176.07	.3	.2	.2
ON HAND ADJUSTMENTS						
INCREASES OF ON HAND	27	8	56.97	.1	.1	.1
DECREASES OF ON HAND	18	8	37.98	.0	.1	.0
INVENTORY APPRECIATION						
NEW PRICES	36	20	73.80	.1	.3	.1
CHANGES PRICES	43652	7860	14103.00	100.0	100.0	14.0
PART MOVEMENT ANALYSIS						
MOVEMENT IN						
0 THRU 3 MONTHS	18789	3334	34525.05	43.1	42.4	34.2
4 THRU 6 MONTHS	10031	1859	25343.29	23.0	23.7	25.1
7 THRU 12 MONTHS	8427	1467	22217.27	19.3	18.7	22.0
13 THRU 24 MONTHS	5168	938	13673.21	11.8	11.9	13.5
MORE THAN 24 MONTHS	1232	262	5273.66	2.8	3.3	5.2
STATUS AND CLASS CODE TOTALS						
AUTOMATIC PHASEOUTS	316	72	612.86	.7	.9	.6
DEALER PHASEOUTS	172	56	384.60	.4	.7	.4
TEMPORARY STOCKS	61	57	294.43	.1	.7	.3
REPLACES PARTS	2167	785	4315.81	5.0	10.0	4.3
NON-RETURNABLE PARTS	4263	1675	8631.27	9.8	21.3	8.5
FROZEN MIN/MAX ANALYSIS						
FROZEN MINIMUM SHORTAGES	180	36	416.42	.4	.5	.4
FROZEN MAXIMUM SHORTAGES	216	42	176.21	.5	.5	.2
FROZEN MINIMUM SURPLUSES	86	18	43.13	.2	.2	.0
FROZEN MAXIMUM SURPLUSES	95	21	191.18	.2	.3	.2
PERCENTAGE OF FILL						
STOCK CUSTOMER FILL LEVEL				92.1	88.6	90.8
CUSTOMER FILL LEVEL				80.5	73.2	71.2

PAGE   1

MELTON EQUIPMENT CO.
321 MAIN STREET
CENTER CITY, U.S.A.

000208   21200999

A & F FARM SERVICE
ROUTE 1 BOX 10
ABBEVILLE AL   36310

STATEMENT

1-31-77
MO.   DAY   YR.
BILLING DATE

Payments, credits or charges received after the billing date shown above will appear on your next statement.

PLEASE DETACH UPPER PART OF STATEMENT AND RETURN WITH REMITTANCE.

FINANCE CHARGE is computed by periodic rate of **1.00** % per month which is ANNUAL PERCENTAGE RATE of **12.00** % applied to all charges which have become more than 30-days past due as of billing date shown on this statement. Minimum finance charge   **.00**

DATE MO. DAY		DESCRIPTION	CHARGES	PAYMENTS AND CREDITS	NEW BALANCE
		PREVIOUS BALANCE	196.45		
1-02	10	01017		96.45	
1-09	10	01096	5.61		
1-18	10	02054	26.90		
1-28	10	02058	142.49		
FINANCE CHARGE			1.00		

ANALYSIS OF ACCOUNT

CURRENT	30-DAYS	60-DAYS & OVER	PLEASE PAY THIS AMOUNT	
175.00	.00	100.00		276.00

TO AVOID ADDITIONAL FINANCE CHARGE PAY NEW BALANCE BEFORE NEXT BILLING DATE

DEALER 7-99-9999-9   NFPEDA TEST DEALER

| CUST. NO. | NAME | BALANCE | CURRENT | AGED | | | SERVICE CHARGE | LAST PD DATE | LAST PD AMOUNT |
| | | | | ***** C E B I T   B A L A N C E S ***** | | | | | |
				1 MONTH	2 MONTHS	3 MONTHS			
18	LARRY ARMSTRONG	11,187.62	1,000.00	3,000.00	3,000.00	4,C61.85	125.77	5/31/79	5.68
91	AUGUSTA FARM SUPPLY	1,186.96				1,172.31	14.65		.00
216	BILL BAILEY	129,536.31	13,456.06	40,368.18	40,368.18	33,S1C.81	1,433.08	5/31/79	500.00
299	BAPTIST HOSPITAL EAST	240.75				237.78	2.97		.00
356	ROBERT DAHL	185.68				183.39	2.29	1/24/79	100.00
414	MABEL DRURY	1,815.28CR	1,815.28CR					5/31/79	207.92
489	RAY GARDINER	202.31				199.82	2.49		.00
547	GUARDIAN ANGEL CHURCH	180.47				178.25	2.22		.00
679	HOBBS FARM	1,856.76				1,833.84	22.92		.00
737	HERMAN HODEL	433.82				428.47	5.35		.00
943	KENTUCKY HIGHWAY DEPT	223.58				220.82	2.76		.00
1008	GILBERT NAY	108.79				107.45	1.34		.00
1099	EDWARD PHILPOT	1,198.30				1,183.51	14.79		.00
1172	WILLIAM SCHMIDT	27.75				27.25	.50		.00
1255	DR DAVID SMILEY	2,863.49				2,828.14	35.35		.00
1321	JOHN TAPP	175.96				173.79	2.17		.00
1412	TOWN & COUNTRY ALTO SUPPLY	5,012.58				4,S5C.7C	61.88		.00
1529	RUBIN WATSON	2,333.71CR	2,333.71CR					5/31/79	1,000.00
1669	BOB WRIGHT	188.13				185.81	2.32		.00
1784	RODNEY ZINK	6,025.11	600.00	1,800.00	1,800.00	1,758.14	66.97	5/31/79	14.43

TOTAL DEBITS	160,834.37						
TOTAL CREDITS	4,148.99						
NET AMOUNT DUE	156,685.38	10,907.07	45,168.18	45,168.18	53,642.13	1,799.82	
	100.0%	7.0%	28.8%	28.8%	34.2%	1.1%	

VENDOR NAME	VENDOR NUMBER *	*	TCTAL DUE *	CURRENT *	30 DAYS *	60 DAYS *	90 DAYS *	120 DAYS *	YTD *PURCHASES *
		*	*	*	*	*	*	*	*
BILL SMITH	CCCC34 *	*	*	*	*	*	*	*	*
SHERRY WILSON	CCC125 *	*	*	*	*	*	*	*	*
ALTMAN TRACTOR & EQUIP	CCC2C8 *	*	2697.37*	2697.37*	*	*	*	*	* 659.63-
AMCO PRODUCTS	CCC4C6 *	*	2136.62*	2136.62*	*	*	*	*	* 1824.60*
AUSTIN PRODUCTS	CCC6C4 *	*	4144.09*	4144.09*	*	*	*	*	* 1063.50-
CCLE MFG CO	CCC8C2 *	*	191.76*	191.76*	*	*	*	*	* 1670.20-
THE CROPPER CO	CC10C8 *	*	1872.59-	245.75-	1002.00-	624.84-	*	*	* 2249.75-
M G DICKEY CO	CC12C6 *	*	7586.88*	7586.88*	*	*	*	*	* 2665.75-
GASTON SEALEY CO	CC14C4 *	*	4.24-	*	4.24-	*	*	*	* 4.24-
HESSTON CORP	CC16C2 *	*	6030.51*	6030.51*	*	*	*	*	* 78.46-
IMPLEMENT SALES	CC18C0 *	*	2407.12-	*	1203.56-	1203.56-	*	*	* 2407.12-
LEWIS BROS	CC20C6 *	*	4060.10-	*	2030.05-	2030.05-	*	*	* 4060.10-
LILLISTON CORP	CC22C4 *	*	10178.04-	*	5089.02-	5089.C2-	*	*	* 10178.04-
LCNG MFG CO	CC24C2 *	*	12005.02-	*	6002.51-	6002.51-	*	*	* 12005.02-
LOVE TRACTOR SALES	CC26C0 *	*	1082.20-	*	579.84-	502.36-	*	*	* 1082.20-
CUSLEY BROS	CC2808 *	*	604.00-	*	302.00-	302.00-	*	*	* 604.00-
REYNOLDS RESEARCH & MFG	CC30C4 *	*	55.78-	*	27.89-	27.89-	*	*	* 55.78-
E J SMITH	CC3202 *	*	904.00-	*	452.00-	452.00-	*	*	* 904.00-
SPARROW & FAIR TR CO	CC34C0 *	*	27.57-	*	27.57-	*	*	*	* 27.57-
SPERRY NEW HOLLAND	CC36C8 *	*	1847.12-	*	923.56-	923.56-	*	*	* 1847.12-
SPRAYRITE MFG CO	CC38C6 *	*	37.88-	*	37.88-	*	*	*	* 37.88-
TCDD CO	CC40C2 *	*	216.58-	*	108.29-	108.29-	*	*	* 216.58-
UNVERFERTH MC CURDY	CC42C0 *	*	6107.14*	6107.14*	*	*	*	*	* 1.83-
JCB P WYATT	CC44C8 *	*	2016.21*	2016.21*	*	*	*	*	*
SUSPENSE	555555 *	*	*	*	*	*	*	*	*

JUNE, 1980

SERVICEMAN: EMPLOYEE # 006 LEVEL: 4 DEPT: 4 SERVICE CONS. PG. NO. 174

DEALER NO.

DATE CLSD	REPAIR ORDER	R/O SKL LVL	* * FLAT RATE ACTIVITY * * HRS WKD	HRS BLD	GAIN/ LOSS	% GAIN OR LOSS	* * STRAIGHT TIME ACTIVITY * * HRS WKD	HRS BLD	GAIN/ LOSS	% GAIN OR LOSS	LABOR COSTS	LABOR BILLED	GROSS MARGIN DOLLARS	%
CUSTOMER TIME:														
1	28362	3					3.1	3.1			14.57	77.50	62.93	81.2
18	28452	3	10.9	9.3	1.6-	14.7-	23.6	17.7	5.9-	25.0-	171.37	513.00	341.63	66.6
16	28453	2	7.7	2.1	5.6-	72.7-	6.4	6.4			78.85	161.50	82.65	51.2
16	28454	2					2.1	2.1			15.00	39.90	24.90	62.4
25	28470	3					2.0	2.0			9.40	38.00	28.60	75.3
28	28525	3					7.8	7.8			39.14	166.20	127.06	76.5
6	28608	3	3.2	3.6	.4	12.5	4.1	4.1			40.97	168.90	127.93	75.7
12	28638	3					3.8	3.8			19.57	95.00	75.43	79.4
13	28655	3					4.0	4.0			20.61	100.00	79.39	79.4
14	28661	3					2.3	2.3			11.85	57.50	45.65	79.4
14	28669	3					1.3	1.3			6.70	32.50	25.80	79.4
26	28704	3					.4	.4			2.06	8.40	6.34	75.5
26	28737	3					.9	.9			4.64	22.50	17.86	79.4
28	28746	3					5.1	5.1			26.27	127.50	101.23	79.4
30	28773	3					3.3	3.3			17.00	82.50	65.50	79.4
3	28798	3					2.4	2.4			12.36	84.00	71.64	85.3
4	28800	3					1.7	1.7			8.76	35.70	26.94	75.5
9	28802	3					2.9	2.9			14.94	72.50	57.56	79.4
6	28807	3					3.0	3.0			15.45	75.00	59.55	79.4
5	28813	3					.5	.5			2.58	10.50	7.92	75.4
6	28815	3					3.3	3.3			17.00	82.50	65.50	79.4
9	28827	3					3.2	3.2			16.48	67.20	50.72	75.5
10	28835	3	7.2	8.5	1.3	18.1	3.1	3.1			53.06	246.40	193.34	78.5
11	28846	3					3.9	3.9			20.09	97.50	77.41	79.4
11	28847	3					1.0	2.2	1.2	120.0	5.15	77.00	71.85	93.3
19	28859	3					.3	.3			1.55	6.30	4.75	75.4
25	28865	3					1.3	2.6	1.3	100.0	6.70	65.00	58.30	89.7
19	28866	3					2.0	1.0	1.0-	50.0-	10.31	21.00	10.69	50.9
17	28887	3					.4	.4			2.06	8.40	6.34	75.5
19	28890	3					3.6	3.6			18.54	72.00	53.46	74.3
24	28909	3					.6	.6			3.09	12.60	9.51	75.5
26	28919	3					.9	.9			4.64	18.90	14.26	75.4
TOTAL CUSTOMER			29.0	23.5	5.5-	19.0-	104.3	99.9	4.4-	4.2-	690.76	2,743.40	2,052.64	74.8
INTERNAL:														
29	28231	2					4.0	7.1	3.1	77.5	18.80	139.60	120.80	86.5
3	28559	2					1.1	1.1			5.67	22.00	16.33	74.2
2	28572	2					2.3	2.3			10.81	38.64	27.83	72.0
5	28794	3					4.1	4.1			21.12	68.88	47.76	69.3
11	28805	3		.5	.5		1.0	2.3	1.3	130.0	5.15	54.40	49.25	90.5
13	28862	3					.9		.9	100.0	4.64	0.00	4.64-	0.0
18	28870	3	1.8	2.0	.2	11.1					9.27	33.60	24.33	72.4
20	28894	3					2.9	2.9			14.94	48.72	33.78	69.3
24	28901	3					2.9	2.9			14.94	58.00	43.06	74.2
TOTAL INTERNAL			1.8	2.5	.7	38.9	19.2	22.7	3.5	19.2	105.34	463.84	358.50	77.3

WARRANTY:

DEALER NO.

SERVICEMANS MONTHLY REPORT

PAGE 14

JUNE, 1980

SERVICEMAN: B S   1   EMPLOYEE # 006   LEVEL: 4   DEPT: 4 SERVICE   CONS. PG. NO. 176

RECAP:	THIS MONTH	YEAR TO DATE
CONTRIBUTION	2,529.05	5,565.32
% CONTRIBUTION TO BILLED	74.8	75.4
TOTAL HOURS WORKED	163.2	355.9
TOTAL HOURS BILLED	157.7	337.1
% TOT LABOR HOURS GAIN/LOSS	3.4-	5.3-
CONTRIBUTION PER HOUR WORKED	15.50	15.64

CONTRIBUTION BY SKILL LEVEL OF JOBS PERFORMED
(DOES NOT INCLUDE FLAT RATE BONUS AMOUNT)

R/O SKILL LEVEL	HOURS WORKED	THIS MONTH	MIX %	YEAR TO DATE	MIX %
0	6.7	69.80	2.8	299.42	5.4
1	0.0	0.00	0.0	48.57	0.9
2	22.5	256.18	10.1	1,190.27	21.4
3	134.0	2,203.07	87.1	4,027.06	72.4
4	0.0	0.00	0.0	0.00	0.0
COMEBACK	0.0	0.00	0.0	0.00	0.0
TOTAL	163.2	2,529.05	100.0	5,565.32	100.0
BONUS	0.0	0.00	0.0	0.00	0.0
TOTAL CONTRIB	163.2	2,529.05	100.0	5,565.32	100.0

JANUARY 1982
* * * G R O S S   M A R G I N * * * * * * *

* *CONS. PG. NO. 211N * *

DEPT: 4 SERVICE	F/R HOURS GAIN (LOSS)	F/R % GAIN (LOSS)	S/T HOURS GAIN (LOSS)	CUSTOMER		INTERNAL		WARRANTY		COMEBACK		BONUS	THIS MONTH		YEAR TO DATE	
EMPLOYEE				AMOUNT	% TO BLD.	AMOUNT	% TO BLD.	AMOUNT	% TO BLD.	HOURS	COST		AMOUNT	% TO BLD.	AMOUNT	% TO BLD.
K	4.1	6.7	8.2-	1,704	73.4	1,032	78.5	21	76.1	0	0	155	2,602	71.0	19,188	64.5
E A	13.8-	21.1-	33.9-	552	75.9	452	71.0	416	54.2	0	0	0	1,420	66.6	17,948	64.8
S A	5.6-	9.5-	34.5-	202	42.8	1,227	76.6	195	66.2	0	0	51	1,573	66.4	14,709	62.8
J W	10.3-	7.3-	13.1-	2,692	80.5	46	78.1	162	81.9	0	0	817	2,083	57.8	22,984	64.8
D J	.0	.0	.0	0	.0	0	.0	0	.0	0	0	0	0	.0	106	96.4
B S	.0	.0	.0	0	.0	0	.0	0	.0	0	0	0	0	.0	37-	.0
M S	.0	.0	.0	0	.0	0	.0	0	.0	0	0	22	22-	.0	203	12.5
C J	4.0-	7.6-	5.5-	1,795	81.3	100	92.5	57	79.9	4	23	588	1,341	56.2	23,219	66.0
M	5.5	6.7	14.2-	1,237	75.6	42	67.1	1,082	70.3	1	8	360	1,993	61.6	21,959	65.7
** NOT FOUND **	.0	.0	.0	0	.0	0	.0	0	.0	0	0	0	0	.0	60	84.5
M J	11.6-	15.4-	14.0-	316	66.7	576	72.8	926	74.0	0	0	85	1,733	68.9	13,433	64.9
W	6.9	6.4	3.0-	3,596	74.9	196	73.9	38	75.1	0	0	4	3,826	74.8	24,821	71.2
J	.0	.0	.0	0	.0	0	.0	0	.0	0	0	0	0	.0	2,830	59.9
J	.0	.0	.3	0	.0	0	.0	6	87.0	0	0	0	6	85.7	6,345	81.5
J	5.9-	15.9-	16.3-	225	73.1	741	74.2	506	67.3	0	0	269	1,203	58.4	13,133	61.5
S	2.8	.0	3.3	96-	99.8	0	.0	230	100.0	0	0	0	134	100.0	65	86.7
J	.9	5.7	3.3	425	76.7	548	78.8	180	92.1	0	0	147	1,006	69.6	4,236	66.3
** NOT FOUND **	.0	.0	.0	0	.0	0	.0	0	.0	0	0	0	0	.0	77	100.0
TOTAL	31.0-	4.4-	135.8-	12,648	75.5	4,960	75.9	3,819	70.9	5	31	2,498	18,898	65.9	185,279	65.5
PLAN VARIANCE				7,284		848-		440			21	2,498-				

DESCRIPTION	CURRENT MONTH AMOUNT	% TO SALES	VAR FROM PLAN	YEAR TO DATE AMOUNT	% TO SALES	VAR FROM PLAN	DESCR	LABOR SALES HOURS	VAR HRS	AMOUNT	VARIANCE AMOUNT
GROSS MARGIN (CONTRIBUTION + COMEBACK)	18,881	64.9	3,883	186,026	65.9	22,977-	MONTH	1,260	1,260	29,820	7,780
							YTD	12,722	12,722	296,167	11,770
OTHER SOURCES OF GROSS MARGIN MACHINE EARNINGS	1,153	100.0	706	13,436	100.0	3,014					

DEALER NO.   S E R V I C E   D E P A R T M E N T   M O N T H L Y   R E P O R T   PAGE 2

JANUARY 1982

CONS. PG. NO. 212

DESCRIPTION	- - CURRENT MONTH - -			* * YEAR TO DATE * *		
	AMOUNT	% TO SALES	VAR FROM PLAN	AMOUNT	% TO SALES	VAR FROM PLAN
OUTSIDE LABOR & MAT	83	33.47	17	1,388	29.97	465
TOTAL MARGIN	20,117	66.0	5,053	200,850	66.9	9,076-
DIRECT CONTROLLABLE EXPENSE						
TRAINING	304	1.00	256-	1,797	.60	1,147-
ADVERTISING & PROMOTION	0	.00	68-	0	.00	320-
AFTER SALES EXP-COMEBACK	32	.10	22-	9,636	3.21	6,466-
SERVICE SALARIES	2,036	6.67	394	36,837	12.26	558
SERVICEMEN UNAPPLIED TIME	0	.00	2	0	.00	280
SHOP SUPP & EQUIP REPAIRS	3,117	10.22	897-	23,482	7.82	8,662-
OTHER CARS & TRUCK EXP	741	2.43	963	16,712	5.56	928
JANITORIAL SERV	361	1.18	113	4,485	1.49	45
TOTAL DIRECT CONTROLLABLE	6,591	21.6	229	92,949	30.9	14,784-
CONTROLLABLE MARGIN	13,526	44.3	5,282	107,901	35.9	23,860-
DIRECT UNCONTROLLABLE EXPENSE						
AFTER SALES EXPENSES	2,200	7.21	2,044-	19,175	6.38	13,835-
OFFICE SUPPLIES & EXPENSE	0	.00	4	136	.05	54-
POSTAGE	0	.00		2	.00	2-
EMPLOYEE FRINGE BENEFITS	541	1.77	347-	1,841	.61	407-
FICA	983	3.22	34-	11,956	3.98	1,712-
STATE UNEMPLOYMENT TAX	0	.00	73	588	.20	287
FEDERAL UNEMPLOYMENT TAX	0	.00	42	305	.10	35
GROUP INSURANCE (H&W)	812	2.66	38-	6,924	2.31	2,064-
WORKMEN'S COMP. INSURANCE	1,071	3.51	439-	7,500	2.50	2,436-
VACATION & HOLIDAY PAY	288	.94	892	8,751	2.91	69
WORKMANS COMP SHOP	0	.00		136	.05	136-
EMP. PROFIT SHARING PLAN	111	.36	209	1,351	.45	2,819
INS-OTHER THAN REAL EST	261	.86	24	1,095	.36	1,420
INSURANCE-REAL ESTATE	357	1.17	39-	2,500	.83	38
BONUS-EMPLOYEES-N/S	0	.00		4,050	1.35	1,950
TOTAL DIRECT UNCONTROLLABLE	6,624	21.7	1,697-	66,310	22.1	14,028-
CONTRIBUTION MARGIN	6,902	22.6	3,585	41,591	13.9	37,888-

ANALYSIS-LABOR HOURS WORKED

DESCRIPTION	CUR MO	MIX %	YTD	MIX %
SALES	1422.1	67.9	13,948.7	65.1
COMEBACK	5.6	.3	307.2	1.4
NONSALEABLE	666.0	31.8	7,186.6	33.5
UNASSIGNED	.0		.0	
TOTAL	2093.7	100.0	21,442.5	100.0

LABOR RECOVERY

CURR MO   60.2%

YTD   59.3%

WORK IN PROCESS

	HOURS	AMOUNT
CURR MO	648	3,143
PRIOR MO	735	3,643

A6900R01   R E P O R T   C F   S I G N I F I C A N T   R A T I O S   DATE CF RUN C8/12/81
DEALER 6-06-0001-6-01   AS CF 09/30/81

                    RATICS ARE PERCENTAGE FIGURES CF DOLLAR CCMPARISCNS. THEY REFLECT
                    SYMPTCMS CNLY. NCT SPECIFIC PRCBLEMS. TC BE MEANINGFUL THEY MUST BE
                    MCNITCRED CVER A PERICD CF TIME AND CCMPARED WITH OTHER DEALERS

NET WORKING CAPITAL TURNOVER IS CALCULATED BY DIVIDING   INVENTCRY TO NET WORKING CAPITAL IS CALCULATED BY DIVIDING
SALES BY NET WORKING CAPITAL   INVENTORY BY NET WORKING CAPITAL

    AS OF   THIS YR   LAST YR   AS OF   THIS YR   LAST YR

    FEB   .00 TU 1   4.10 TC 1   FEB   .00 TO 1   1.59 TC 1
    MAR   .00 TO 1   3.14 TC 1   MAR   .00 TO 1   3.8C TC 1
    APR   .00 TO 1   1.14 TC 1   APR   .50 TO 1   1.97 TC 1
    MAY   .10 TO 1   .94 TC 1   MAY   .23 TO 1   3.98 TC 1
    JUN   .00 TO 1   2.06 TC 1   JUN   .09 TO 1   4.51 TC 1
    JUL   .00 TO 1   .58 TC 1   JUL   .09 TO 1   1.90 TC 1
    AUG   .05 TO 1   1.85 TC 1   AUG   .46 TO 1   1.37 TC 1
    SEP   .00 TO 1   1.81 TC 1   SEP   .87 TO 1   .95 TC 1
    OCT   2.92 TC 1   OCT   .8C TC 1
    NOV   7.60 TC 1   NOV   1.89 TC 1
    DEC   4.26 TC 1   DEC   .91 TC 1
    JAN   3.57 TC 1   JAN   18.40 TC 1

TURNOVER OF TOTAL ASSETS IS CALCULATED BY DIVIDING SALES BY   RETURN ON ASSETS IS CALCULATED BY DIVIDING NET PRCFIT BY
TOTAL ASSETS   TOTAL ASSETS

    AS OF   THIS YR   LAST YR   AS OF   THIS YR   LAST YR

    FEB   .00 TO 1   3.86 TC 1   FEB   .00 TO 1   .35 TC 1
    MAR   .00 TO 1   2.94 TC 1   MAR   .00 TO 1   .37 TC 1
    APR   .00 TO 1   1.C7 TC 1   APR   .06 TO 1   .C2 TC 1
    MAY   4.10 TO 1   .79 TC 1   MAY   8.38 TO 1   .C5 TC 1
    JUN   .07 TO 1   1.37 TC 1   JUN   2.33 TO 1   .C9 TC 1
    JUL   .00 TO 1   .41 TC 1   JUL   .00 TO 1   .C3 TC 1
    AUG   .21 TO 1   .90 TC 1   AUG   .00 TO 1   .C4 TC 1
    SEP   .03 TO 1   .72 TC 1   SEP   .75 TO 1   .2C TC 1
    OCT   .82 TC 1   OCT   .C4 TC 1
    NOV   2.C3 TC 1   NOV   .11 TC 1
    DEC   1.63 TO 1   DEC   .12 TO 1
    JAN   1.00 TC 1   JAN   .07 TO 1

DATE	CHK #	GROSS	FED W/H	FICA	STATE W/H	CITY W/H	ACCT REC EMPL	INS ACCR	WITHDRWL BF	WITHDRWL EF	WITHDRWL TS	NET PAY	ERROR

***** JOHN SMITH   # C24CCC   SCC SEC # 123-44-5678

DATE	CHK #	GROSS	FED W/H	FICA	STATE W/H	CITY W/H	ACCT REC EMPL	INS ACCR	WITHDRWL BF	WITHDRWL EF	WITHDRWL TS	NET PAY	ERROR
CLC Q-T-D BAL		4036.92	327.50-	270.15-	72.78-	.00	381.31-	.00	.CC	.CC	.CC	2985.18-	
OLD Y-T-D BAL		4036.92	327.50-	270.15-	72.78-	.00	381.31-	.00	.0C	.0C	.00	2985.18-	
09/15	513	1444.00	1CC.CC-	2CC.CC-	44.00-	.00	.00	.00	.00	.CC	.00	1100.00-	
TCTAL THIS MO		1444.00	1CC.CC-	2CC.GC-	44.00-	.00	.00	.C0	.CC	.CC	.CO	11CC.00-	
NEW Q-T-D BAL		5480.92	427.50-	470.15-	116.78-	.00	381.31-	.00	.00	.0C	.00	4085.18-	
NEW Y-T-D BAL		5480.92	427.50-	470.15-	116.78-	.00	381.31-	.00	.CC	.0C	.0C	4085.18-	

***** SHERRY ERKER   # C242G8   SCC SEC # 248-68-3238

DATE	CHK #	GROSS	FED W/H	FICA	STATE W/H	CITY W/H	ACCT REC EMPL	INS ACCR	WITHDRWL BF	WITHDRWL EF	WITHDRWL TS	NET PAY	ERROR
CLC Q-T-D BAL		550.68	55.9C-	37.58-	17.81-	.00	10.00-	.00	.0C	.0C	.00	429.39-	
CLD Y-T-D BAL		550.68	55.9C-	37.58-	17.81-	.00	10.00-	.00	.0C	.CC	.0C	429.39-	
TCTAL THIS MO		.CO	.CC	.CC	.00	.00	.00	.C0	.CC	.CC	.CC	.00	
NEW Q-T-D BAL		550.68	55.9C-	37.58-	17.81-	.00	10.00-	.00	.0C	.CC	.00	429.39-	
NEW Y-T-D BAL		550.68	55.9C-	37.58-	17.81-	.00	10.00-	.00	.0C	.CC	.00	429.39-	

***** JOHN WELLS   # C25874   SOC SEC # 247-82-3777

DATE	CHK #	GROSS	FED W/H	FICA	STATE W/H	CITY W/H	ACCT REC EMPL	INS ACCR	WITHDRWL BF	WITHDRWL EF	WITHDRWL TS	NET PAY	ERROR
CLC Q-T-D BAL		2939.57	343.6C-	199.55-	77.25-	.00	3.14-	.00	.0C	.CC	.00	2316.03-	
OLD Y-T-D BAL		2939.57	343.6C-	199.55-	77.25-	.00	3.14-	.00	.00	.0C	.00	2316.03-	
TCTAL THIS MO		.00	.CC	.00	.00	.00	.00	.C0	.CC	.CC	.0C	.00	
NEW Q-T-D BAL		2939.57	343.6C-	199.55-	77.25-	.00	3.14-	.00	.0C	.CC	.00	2316.03-	
NEW Y-T-D BAL		2939.57	343.6C-	199.55-	77.25-	.00	3.14-	.00	.0C	.CC	.00	2316.03-	

***** CHARLES P FLESCHNER   # C27466   SCC SEC # 250-54-3795

DATE	CHK #	GROSS	FED W/H	FICA	STATE W/H	CITY W/H	ACCT REC EMPL	INS ACCR	WITHDRWL BF	WITHDRWL EF	WITHDRWL TS	NET PAY	ERROR
CLF Q-T-D BAL		3832.41	46C.30-	261.42-	117.87-	.00	.00	.C0	.CC	.CC	.00	2592.82-	
CLD Y-T-D BAL		3832.41	46C.30-	261.42-	117.87-	.00	.00	.00	.00	.0C	.00	2592.82-	
TCTAL THIS MO		.CO	.CC	.CC	.00	.00	.00	.C0	.CC	.CC	.00	.00	
NEW Q-T-D BAL		3832.41	46C.30-	261.42-	117.87-	.00	.00	.00	.CC	.CC	.0C	2592.82-	
NEW Y-T-D BAL		3832.41	46C.30-	261.42-	117.87-	.00	.00	.00	.00	.0C	.00	2592.82-	

***** DAVE FORMENTI   # C2886C   SCC SEC # 248-74-7658

DATE	CHK #	GROSS	FED W/H	FICA	STATE W/H	CITY W/H	ACCT REC EMPL	INS ACCR	WITHDRWL BF	WITHDRWL EF	WITHDRWL TS	NET PAY	ERROR
OLD Q-T-D BAL		.00	.00	.00	.00	.00	.00	.00	.0C	.CC	.00	.00	
OLD Y-T-D BAL		.00	.CC	.CC	.00	.00	.00	.00	.00	.CC	.CC	.00	
TCTAL THIS MO		.00	.CC	.00	.00	.00	.00	.C0	.00	.CC	.00	.00	
NEW Q-T-D BAL		.00	.00	.CC	.00	.00	.00	.00	.00	.CC	.0C	.00	
NEW Y-T-D BAL		.00	.CC	.00	.00	.00	.00	.00	.0C	.CC	.00	.00	

An Accounting Problem

The successful manager must know the rudiments of accounting. Five basic rules govern the transactions in double-entry accounting. Review the rules, post the transactions on the Balance Sheet Accounts and Profit and Loss Accounts complete the Trial Balance, Profit and Loss Statement and Balance Sheet. Familiarity with this procedure will smooth your communications with your bookkeeper.

THE FIVE BASIC RULES THAT GOVERN ALL TRANSACTIONS IN DOUBLE-ENTRY ACCOUNTING

1. When assets are increased, their accounts are debited. When they are decreased, their accounts are credited.

2. When liabilities increase, their accounts are credited. When they are decreased, their accounts are debited.

3. When net worth is increased, its accounts are credited. When it is decreased, its accounts are debited.

4. When income is increased, its accounts are credited. When income is decreased, its accounts are debited.

5. When expenses are increased, their accounts are debited. When they are decreased, they are credited.

The Five Basic Rules Diagram

RULE	TYPE	DEBIT	CREDIT
1	Assets	+	−
2	Liabilities	−	+
3	Net Worth	−	+
4	Income	−	+
5	Expense	+	−

SAMPLE TRACTORS, INC.

OPENING BALANCE SHEET

Assets

Cash	$ 30,000
Receivables	90,000
Inventories	500,000
Fixed Assets	50,000
Total Assets	**$670,000**

Liabilities & Net Worth

Accounts Payable & Accruals	$150,000
Floor Plan Payables	320,000
Notes Payable	60,000
Mortgages Payable	30,000
Net Worth	110,000
Total Liabilities & Net Worth	**$670,000**

SAMPLE TRACTORS, INC.

LIST OF TRANSACTIONS

1. Received four carloads of equipment from manufacturer—$160,000.

2. Sold new tractor, (list price $40,000), took used tractor in trade ("as is" value $18,000), and received cash of $16,000.

3. Paid manufacturer for above transaction—$31,000.

4. Borrowed money from bank—$10,000.

5. Paid accounts payable—$3,000.

6. Recorded Service Parts sales for month—$40,000. Cash received for the above sales was $10,000. The remaining sales were charge sales.

7. Recorded Cost of Sales for the Service Parts transaction above. The Gross Profit rate for service parts is 25%.

8. The mechanics' wages for the month were $3,000. Customer labor charges amounted to $4,000, and internal labor charges at retail totaled $2,000.

9. Recorded depreciation for the period—$1,000.

10. Bought new pick-up truck—$7,000.

11. Made mortgage payment of $3,000.

12. Recorded salaries and other operating expenses for the period of $9,000.

13. Prepare Trial Balance.

14. Determine profit for the period.

15. Prepare closing entries for the period.

16. Prepare Balance Sheet and Profit and Loss Statement for the period.

SAMPLE TRACTORS, INC.

BALANCE SHEET ACCOUNTS

CASH		RECEIVABLES		INVENTORIES	
30,000		90,000		500,000	

FIXED ASSETS—NET		ACCOUNTS PAYABLE		FLOOR PLAN PAYABLES	
50,000			150,000		320,000
					160,000

NOTES PAYABLE		MORTGAGE PAYABLE		NET WORTH	
	60,000		30,000		110,000

SAMPLE TRACTORS, INC.

P & L ACCOUNTS

SALES	COST OF SALES

OVER-ALLOWANCES	SALARIES AND OPERATING EXPENSES

DEPRECIATION

SAMPLE TRACTORS, INC.

TRIAL BALANCE

Balance Sheet
  Cash _______________________
  Receivables _______________________
  Inventories _______________________
  Fixed Assets _______________________
  Accounts Payable _______________________
  Floor Plan Payables _______________________
  Notes Payable _______________________
  Mortgage Payable _______________________
  Net Worth _______________________

Profit and Loss Statement
  Sales _______________________
  Cost of Sales _______________________
  Overallowances _______________________
  Salaries and Operating Expenses _______________________
  Depreciation _______________________

SAMPLE TRACTORS, INC.

Profit and Loss Statement _______________________
  Sales _______________________
  Cost of Sales _______________________
  Overallowances _______________________
  Gross Profit _______________________

Less Expenses:

  Salaries & Operating Expenses
  Depreciation _______________________

Net Profit for Period _______________________

Balance Sheet   *Liabilities and Net Worth*
Assets
  Current Assets   Current Liabilities
  Cash _____________   Accounts Payable _____________
  Receivables _____________   Floor Plan Payables _____________
  Inventories _____________   Notes Payable _____________

  Total _____________   Total _____________

  Fixed Assets - Net _____________   Mortgage Payable _____________

  Total Assets _____________   Net Worth _____________

                              Total Liabilities &
                                Net Worth _____________

Index